Innovation in Japan

The Japanese economy has made a remarkable recovery from the so-called 'Lost Decade' of the 1990s. This said, demographic trends suggest that Japan will have to show remarkable powers of innovation if it is to continue to prosper in the global economy. For, around the turn of the last century texts published by prominent strategy analysts such as Michael Porter and colleagues were asking whether Japan could continue to compete at all, and in answering this question they not only gained significant global attention, they also appeared to sound the death knell for strategic innovation in Japan.

This collection helps put the record straight. It invites authors and editors of previous (Routledge) titles on the topic of 'Innovation in Japan' to reflect on how things have moved on – prominent scholars on Japanese innovation such as Martin Hemmert, Cornelia Storz, and Ruth Taplin, all of whom appear in this collection. It brings together fresh perspectives on Japanese-style innovation, from insiders and from outsiders, from scholars and from practitioners, all of whose combined contributions to this book update our understanding of how patterns of innovation in Japan are evolving and thus provide inspiration and guidance for managers and innovators worldwide.

Keith Jackson is a tutor and researcher at The School of Oriental & African Studies (SOAS) – University of London. He also works freelance as an HRM consultant in Europe and Asia. He is book reviews editor of the Taylor & Francis journal 'Asia Pacific Business Review' (APBR) and co-editor of the Routledge 'Working in Asia' series. Recent publications include 'The Changing Face of Japanese Management' (2004, Routledge).

Philippe Debroux is Professor of International Management at Soka University, Japan. He is also international reviews editor of the 'Asia Pacific Business Review' (APBR), one of several journals in which he publishes regularly. Recent publications include 'Human Resource Management in Japan: Changes and Uncertainties: A New Human Resource Management System Fitting to the Global Economy' (2003, Ashgate). He is currently conducting research into female entrepreneurship in Japan.

Innovation in Japan

Emerging Patterns, Enduring Myths

Edited by
Keith Jackson and Philippe Debroux

Routledge
Taylor & Francis Group

LONDON AND NEW YORK

First published 2009
by Routledge
2 Park Square, Milton Park, Abingdon, Oxfordshire OX14 4RN

Simultaneously published in the USA and Canada
by Routledge
711 Third Avenue, New York, NY 10017

Routledge is an imprint of the Taylor and Francis Group, an informa business

First issued in paperback 2015

© 2009 Edited by Keith Jackson and Philippe Debroux

Typeset in Times 11/12pt by Alden Prepress Ltd, Northampton, UK

British Library Cataloguing in Publication Data
A catalogue record for this book is available from the British Library

ISBN 978-0-415-44579-5 (hbk)
ISBN 978-1-138-97277-3 (pbk)

CONTENTS

1   **Innovation in Japan: An Introduction**
KEITH JACKSON & PHILIPPE DEBROUX   1

2   **Innovation Management of Japanese and Korean Firms: A Comparative Analysis**
MARTIN HEMMERT   8

3   **Comparing National Innovation Systems in Japan and the United States: Push, Pull, Drag and Jump Factors in the Development of New Technology**
KATHRYN IBATA-ARENS   30

4   **Growing R&D Collaboration of Japanese Firms and Policy Implications for Reforming the National Innovation System**
KAZUYUKI MOTOHASHI   54

5   **Japanese Intellectual Property and Employee Rights to Compensation**
RUTH TAPLIN   77

6   **From Vertical to Horizontal Inter-Firm Cooperation: Dynamic Innovation in Japan's Semiconductor Industry**
YOSHITAKA OKADA   93

7   **Innovation, Institutions and Entrepreneurs: The Case of 'Cool Japan'**
CORNELIA STORZ   115

8   **The Expected Roles of Business Angels in Seed/Early Stage University Spin-offs in Japan: Can Business Angels act as Saviours?**
MASANOBU TSUKAGOSHI   139

9   **Innovation in Japan: What Role for University Spin-offs?**
PHILIPPE DEBROUX   157

10   **Emerging Patterns and Enduring Myths of Innovation in Japan: Concluding Thoughts**
KEITH JACKSON & PHILIPPE DEBROUX   175

Index   183

Innovation in Japan: An Introduction

KEITH JACKSON & PHILIPPE DEBROUX

Introduction

Because of the depth and scope of its innovation system, Japan continues to set standards that other countries seek to match in their efforts to manage innovation. Since the 1980s this has been true in the field of manufacturing generally and of product and process innovation in particular. Reference to the traditions of innovation in Japan invokes benchmark concepts such as the Toyota Production System (TPS) together with quality circles, Just In Time (JIT) inventory management, industrial robotics combined with a zero defect approach to quality control, a restless philosophy of continuous improvement (*kaizen*) combined (more recently) with a distinctive 'middle-up-down' management approach towards knowledge management in organizations (cf. Nonaka & Takeuchi, 1995; Bird, 2002; Takeuchi & Nonaka, 2004; Trott, 2008).

In this special issue we bring together a collection of recently researched studies that, in combination, explore the extent to which these established features and patterns of 'Japanese-style' innovation and innovation management remain valid. We seek to highlight and explain some of the emerging patterns for innovation in Japan as this still powerful economy seeks a coherent response to the emerging global challenges to its previous and distinctive pre-eminence across a range of business sectors.

For we believe that the study of what is happening in Japan is important in support of developing a critical understanding of the challenge of global competition and in improving one's grasp of the structural, cultural and political projects underlying successful or unsuccessful innovations across the Asia-Pacific region. All developed economies face competition from emerging economies and are experiencing a shift to a knowledge economy in which intangible assets become more valuable. Japan also faces issues generated by a rapidly ageing population and, concomitantly a shrinking domestic market for certain key products and services.

The Role of Central Government

In Japan, the state has been the traditional driving force for innovation, and above all in the less developed regions of the country. In the case of Japan, strong traditional connections between the political world and large companies often lead to decisions that are skewed in favour of the large corporations and to the detriment of other, smaller companies. This is an enduring reality of business and political life in Japan.

Even during the long period of restructuring at public and private levels, Japan has pursued a policy of innovation planning more aggressively than any other country. Although many people thought that Japan would not have the financial means to compete on the world markets as in the past, the Japanese government has responded with significant investments in the scientific infrastructure, allocation of strong incentives to private R&D and an active support of the commercialization of inventions. Japan remains at the top in terms of R&D expenditure and recent OECD Figures (OECD, 2005/6) show the impressive results of that investment in terms of the number of invention patents, Japan accounting for approximately 25 per cent of the world's total. The question remains, however, as to how effective all this investment has been in securing Japan's position as a major innovative nation.

The Role of Universities and Other Non-governmental Agencies

As several studies presented in this current collection demonstrate, Japanese universities and research centres are being challenged to adopt new and more proactive roles in helping to transform the nation's innovation system. New forms of business relationships are being negotiated between organizations and other agencies operating in both the for-profit and not-for-profit sectors. We explore the emerging roles of venture capitalists and business angels, developers of science and technology and other stakeholders whose influence and interests serve to fashion the current and emergent patterns of innovation in Japan. We recognize that Japan is not unique in this endeavour. However, and given the distinctive nature of the country's innovation system hitherto, the starting point for its response to global change and opportunity is specific and worthy of detailed and up-to-date study.

Research Questions

In this collection of specially researched studies we explore the extent to which the corporate response in Japan to the challenges generated by the evolution in science and technology has been similar or different to that evinced in competing Asia-Pacific economies, and why/why not. Exploring such issues serves to give a broad strategic context for the studies presented here.

More specifically, some of the key research questions addressed in this special issue include:

- What does innovation mean in the context of Japanese business?
- What are the traditional or enduring patterns of innovation that make innovation in Japan so distinctive in comparison to other Asia-Pacific nations and economies?
- What are – or appear to be – the emerging patterns of innovation in Japan, and why are these patterns taking this form?
- Who are the key actors engaged in shaping the emerging patterns of innovation in Japan, how, and why?

Context: Innovation in Twenty-first-Century Japan

Ten years ago, Hemmert and Oberländer concluded their prognosis for the state of 'technology and innovation' in twenty-first-century Japan in the following terms:

> The Japanese innovation system appears to follow a pragmatic pattern of development: old, existing organizational units and processes persist, while new ones with sometimes even radically different approaches are added, resulting in the coexistence of 'old' and 'new', 'traditional' and 'innovative' organizational processes (Hemmert & Oberländer, 1998: 17).

At the time, the authors believed that it was difficult to assess whether innovation in Japan would follow the path of fundamental change, gradual evolution, or stagnation. The stagnation option suggested that the patterns established to sustain a (spectacularly successful) 'catch-up' economy would endure in the form of a 'path dependency' focused mainly on the past (cf. Storz, 2006). To critical outsiders, this path might suggest a propensity to wallow in the myth of national innovation supremacy.

By asking bluntly 'Can Japan Compete?', Michael Porter and his respected Japanese colleagues (Porter *et al.*, 2000) offered a benchmark analysis whose influence persists to this day. Despite being anchored in the dark data emerging out of the 1990s or the so-called 'lost decade' of economic activity in Japan, it hangs like a cloud still over the libraries of universities, business schools and government-related institutions in Japan. To illustrate: since the 1950s the mission of the Japan External Trade Organization (JETRO) had been to promote Japanese exports – driven by the type of the type of product and process innovation highlighted above. Their mission now is to promote Japan as an investment opportunity and as a place where foreign investment and business ideas might find a home.[1] The tide, it appears, has turned.

The provocative millennial analysis offered by Porter and others acted as a spur to reassess traditional strengths and develop new patterns in the course of innovation in Japan. The Japanese government has begun a series of policy and legal changes designed to drive on and support new patterns of innovation activity, as illustrated in several studies in this collection, established patterns of business relationships in Japan are being supplanted by new forms of dynamic cooperation and partnership of a style more instantly familiar to Europeans and North Americans (cf. Taplin, 2007). And yet, there remains something distinctive about innovation in Japan.

Summary of the Contributions

This special issue of *Asia Pacific Business Review* allows us to trace the emerging patterns of innovation in Japan as relevant to innovation management research and practice in the twenty-first century. It is appropriate, therefore, that we hear first from Martin Hemmert, whose speculation with Christian Oberländer about what 'technology and innovation in 21st century Japan' might look like we cited above.

In his updated study for this current collection Hemmert makes a timely comparison between emerging patterns of innovation in South Korea and those currently apparent in Japan. He makes a comparative analysis of corporate and management responses to the challenges – and opportunities – generated by the evolution of science and technology in terms of managing research and development (R&D) and risk taking generally in the context of innovation management. He also compares and explains differences in the approach to managing human resources in each national innovation context – a theme we revisit in our concluding discussion. According to Hemmert, one reason why (South) Korean companies have been able to progress more rapidly than Japan in some fields is because they have been able to devise human resource management processes that induce more creativity and higher motivation, and this from a position of crisis: for example, the humiliation of the International Monetary Fund (IMF) intervention in 1997. Hemmert explains how there has been genuine reform in how innovation is financed in South Korea together with an enhanced willingness to experiment with established structures and processes of innovation. What might the drivers of innovation in Japan learn from this example?

The study by Kathryn Ibata-Arens develops this comparative theme by analysing the relative strengths and weaknesses of the national innovation systems (NIS) in the United States and Japan in supporting sustained new business creation, focusing particularly on technology licensing organizations (TLOs) in the life science industries. Drawing on a previous and substantive corpus of research (cf. Ibata-Arens, 2005), she draws comparisons between the ways in which actors in innovation in the United States and in Japan differ in their approaches towards innovation and entrepreneurship; more specifically, how key actors in each national context respond to the combination of 'push, pull, drag and jump' factors that serve to describe contexts for the strategic development of new technology. In common with other contributors to this special issue, Ibata-Arens offers evidence that central government institutions in Japan are working systematically towards promoting a more 'pro-entrepreneurial culture'; or, at least, are now responding more systemically to the challenges that such an emerging culture generates. How might established patterns of big firm innovation in Japan adapt from US models in this regard?

The following contribution is by Kazuyuki Motohashi, who updates recent research (cf. Motohashi, 2005) in order to remind us of how emerging patterns of innovation in Japan are informed by the increasingly significant role of Japanese small and medium-sized enterprises (SMEs). Motohashi explains how SMEs in the form of new technology-based firms (NTBFs) are responding to national innovation policy reforms and the restructuring of the Japanese NIS by forging new and productive R&D collaborations across a range of business sectors: for example, in partnership with universities. Motohashi explains the emerging role that Japanese NTBFs are taking, for example, in their approach to risk management. He argues that NTBFs are becoming not only a potential source of business growth but also a driving force in the reform of the Japanese innovation system.

The contribution by Ruth Taplin develops our understanding of the policy changes affecting the emerging patterns of innovation in Japan, emphasizing how shifts in the nature and expectations defining Intellectual Property (IP) and Employee Rights to Compensation are serving to generate an emerging wave of creativity and in the form of new collaborations between employees and their employers and between inventors and the legal structure that hitherto characterized the status of IP in Japan's NIS. Taplin demonstrates with in-depth insight how the established structures and cultures of business and political life in Japan are being challenged by 'some of the most radical changes in history to ... [the] assessment of valuing intangible assets and ... attitudes to litigation'. p. 287. She goes on to explain how '[T]he speed of change in the field of intellectual property (IP) is occurring at a furious pace and in some cases surpassing that in western countries'. p. 287. We recognize how emerging patterns of innovation in Japan form part of a global phenomenon (cf. Taplin, 2007).

The contribution by Yoshitaka Okada reminds us of how innovation in Japan might have evolved if senior policy makers together with corporate makers of 'macro' and 'micro' business strategy had adopted certain aspects of a reform path at an earlier date. In developing a detailed case study of the Japanese semiconductor industry, Okada illustrates how patterns of inter-firm cooperation are shifting from a predominantly vertical to a more pronounced horizontal orientation (cf. Okada, 2006). 'Such developments', he argues here, 'though characterized as new, remain an extension of the Japanese institutional inheritance; namely, an emphasis on a strategic mix of cooperation and competition, but with a revised understanding of cooperation' p. 287. These emerging patterns of 'intra- and inter-firm cooperation and interaction' serve to challenge enduring myths – propagated by both Japanese and non-Japanese experts – about the cautiously formulaic nature of risk management in large Japanese corporations or kaisha. As suggested already in this introduction, the stature of the Japanese NIS is immense. As such, it will evolve over time by adapting the new to the old, with (echoing Okada here) a new understanding of 'cooperation' being a key feature of this process.

This perspective on change in the Japanese NIS is given sophisticated expression in the concept of 'plasticity', developed in the contribution by Cornelia Storz. Adopting a similar approach to Motohashi, Storz emphasizes the role that SMEs can play in driving reform in the Japanese NIS. Her industry-specific focus highlights a global success story: the Japanese game software industry. This choice of case study allows her to identify the emerging potential of 'cool' Japan to arrest the trend identified by Hemmert in this collection: that is of Japan as a net exporter of new technologies rather than (as now) a net importer. Developing on her work examining the 'path dependency' of established innovation patterns in Japan (cf. Storz, 2006), in this current study Storz gives fresh momentum and insight to the view that innovation in sectors such as game software development challenges the relative inertia of many established institutional structures and cultures for innovation in Japan. To quote her directly: 'In terms of broader policy implications, the case of the Japanese game software sector teaches us that politically it may be wise to embrace variety. The variation of a given set of dominant and peripheral institutions opens up options for new combinations'. p. 287. We believe the 'new combinations' emphasized here by Storz echo the 'new understanding of cooperation' stressed previously by Okada.

The contribution by Masanobu Tsukagoshi gives us relevant and timely insights into the work of Japanese business angels and the potential they generate for promoting new patterns of strategic collaboration and innovation management in Japan; notably, in the context of start-up/seed ventures identified as 'university spin-offs' (USOs). In this study – based on his practical experience as a 'business angel' in Japan and in the USA – Tsukagoshi identifies a 'gap' between (to quote him directly) 'the needs of Japanese USOs for non-financial support and the capacity of existing investors to supply such support' p. 287. We believe this contribution adds a critical and practice-based dimension to our understanding of the emerging patterns of innovation identified previously by Ibata-Arens and Motohashi. It furthermore outlines the potential for a strategic response to the types of context-specific changes and trends identified by Taplin.

The final contribution is by Philippe Debroux who develops Tsukagoshi's focus on USOs. We are reminded of key details in the policy context for innovation in Japan. Drawing on his own practical experience in this field, Debroux explains how, in Japan as elsewhere, large investments in science and technology are long-term investments. Results may come (or not) after decades of trial and error. According to Debroux, regulatory changes will not suddenly make Tokyo University or other top Japanese universities the equivalent of Harvard and Stanford as centres of academic and managerial proficiency. The gap in terms of management capability can be expected to become smaller after a while; however, the mindset and the expectations can be expected to remain different. Japanese scientists and managers of technology describe their career and expectations in terms of monetary and non-monetary reward in ways that are distinct from what can be observed in the USA and some European countries. The mobility of highly skilled people is higher than before in Japan, but remains low by US and even European standards – a theme we develop in our concluding contribution to this special issue.

Conclusions

The enduring pre-eminence of the Japanese economy is testimony to the enduring strength of innovation in Japan. However, questions are emerging that cast doubt on whether this endurance can continue to rely on 'home grown' structures, processes or ideas; on whether established institutions and patterns of innovation suffice to sustain both Japan's competitive position in the world and the remarkable standard of living of its citizens at home. Whether or not the Japanese government is able to adjust its policy position on innovation and give room and sustenance to the patterns of innovation and energy that are clearly emerging remains to be seen. We believe that the contributions brought together here give testimony to this energy and to this critical examination of what innovation in Japan currently means, and might mean in future.

Acknowledgements

The editors would like to thank Professors Chris Rowley and Malcolm Warner for their continuous support in guiding us during the editing of this *APBR* special issue. We would also like to express our appreciation for the work and patience of Matthew McCann and his colleagues at Routledge.

Note

[1] See the JETRO homepage on the JETRO at www.jetro.go.jp (accessed 20 March 2008).

References

Bird, A. (Ed.) (2002) *Encyclopaedia of Japanese Business and Management* (London: Routledge).

Hemmert, M. & Oberländer, C. (Eds) (1998) *Technology and Innovation in Japan: Policy and Management for the Twenty-first Century* (London: Routledge).

Ibata-Arens, K. (2005) *Innovation and Entrepreneurship in Japan: Politics, Organizations and High Technology Firms* (Cambridge: Cambridge University Press).

Motohashi, K. (2005) Economic analysis of university-industry collaborations: the role of new technology based firms in Japanese national innovation reform, *Research Policy*, 34(5), pp. 583–594.

Nonaka, I. & Takeuchi, H. (1995) *The Knowledge Creating Company – How Japanese Companies Create the Dynamics of Innovation* (Oxford: Oxford University Press).

OECD (Organization for Economic Cooperation and Development) (2005/6) Analytical Business Enterprise Research and Development database 2005/6 (ANBERD) OECD, Paris. Available at http://www.oecd.org/document/17/0,3343,en_2825_497105_1822033_1_1_1_1,00.html (accessed 20 March 2008).

Okada, Y. (2006) Institutional changes and corporate strategies for survival in the Japanese semiconductor industry, in: Y. Okada, *Struggles for Survival: Institutional and Organizational Changes in Japan's High-Tech Industries*, pp. 105–154 (Tokyo: Springer-Verlag).

Porter, M., Takeuchi, H. & Sakakibara, M. (2000) *Can Japan Compete?* (Cambridge, MA: Perseus).

Storz, C. (Ed.) (2006) *Small Firms and Innovation Policy in Japan* (Abingdon: Routledge).

Takeuchi, H. & Nonaka, I. (2004) *Hitotsubashi on Knowledge Management* (Chichester: Wiley).

Taplin, R. (Ed.) (2007) *Innovation and Business Partnerships in Japan, Europe and the United States* (Abingdon: Routledge).

Trott, P. (2008) *Innovation Management and New Product Development*, 4th ed. (London: Prentice Hall).

Innovation Management of Japanese and Korean Firms: A Comparative Analysis

MARTIN HEMMERT

Introduction

International interest in the innovation management of Japanese firms has become widespread since they emerged as strong technological competitors to western firms on the world markets in the 1980s. At one time, the strengths of Japanese innovation management were strongly emphasized (for example, see US Department of Commerce, 1990), whereas later on, following a period of stagnation of the Japanese economy in the 1990s, more attention has been given to its potential shortcomings (for example, Porter *et al.*, 2000). Regardless of specific assessments regarding strengths and weaknesses, however, the reference frame of previous studies on Japanese innovation management has predominantly been its comparison with the management of western firms (for example, Okimoto & Nishi, 1994; Wakasugi, 1994; Nonaka & Takeuchi, 1995).

Recently, however, firms from other East Asian countries have also emerged internationally as strong technological competitors. In particular, the achievements of some leading South Korean (subsequently Korean) firms have been of considerable interest. Accordingly, some studies of the innovation management of Korean firms have also been conducted during the last few years (for example, Kim, 1998; Hobday *et al.*, 2004; K. Lee *et al.*, 2005).

Taken together, firms from both countries have shown continuous technological strength and keep challenging their rivals in the global competition game through a large number of product and process innovations. Therefore, an in-depth understanding of their managerial profiles appears to be crucial in order to develop appropriate strategies for competition and collaboration with them.

On the one hand, and given the geographical and cultural proximity between Korea and Japan, it appears reasonable to assume that Japanese and Korean firms are highly similar to each other in their managerial strategies in general and in the field of innovation management in particular. On the other hand, when considering the rather different developmental paths of the Japanese and Korean economies and the dissimilar institutional surroundings under which firms in the two countries are operating, it may rather be expected that there are considerable differences between them. In order to improve our understanding of what is common and what is different between the innovation management style of Japanese and Korean firms, however, an explicit and detailed comparative analysis is needed.

This study intends to contribute to the literature through such a direct comparison, based on a broad review of primary and secondary data on managerial practices in both countries. First, the innovation-related input and output of Japanese and Korean firms will be reviewed on an aggregated level. Thereafter, a comparative review of their innovation management style will be conducted regarding various managerial fields, such as strategic behaviour, technology sourcing, R&D management and human resource management. In particular, what is common and what is different between Japanese and Korean corporate innovation management will be analysed, and also what are the potential strengths and weaknesses of the management styles in both countries. Finally, based on the findings, managerial implications will be derived for Japanese and Korean firms as well as for their international competitors, and directions for further research will be outlined.

Research Questions and Propositions

This study addresses the following research questions:

- What are the similarities and differences between features of corporate innovation management in Japan and in Korea?
- How can these features be explained with historical, institutional and cultural factors?
- What are the resulting potential strengths and weaknesses of innovation management in Japanese and Korean firms?

In addressing these questions, this study advances the following propositions:

- Adopting a comparative perspective, the similarities and differences between innovation management practices in Japan and (South) Korea can be explained with reference to historical, institutional and cultural factors.

- The resulting potential strengths and weaknesses of Japanese and Korean firms in respect of their innovation-related activities can be explained in similar terms.

Context: Comparing Innovation-related Activities by Japanese and Korean Firms

The total research and development (R&D) expenditures of Japanese and Korean firms amounted in 2004, and on a purchasing power parity adjusted base, to US$88.7 billion and US$21.7 billion, respectively (OECD, 2006). Reflecting the relatively bigger size of the Japanese economy, Japanese firms spent roughly four times as much on R&D as Korean firms. When set in relation to each country's gross domestic product (GDP) and to the total value added in industry, however, the R&D intensity of Japanese firms was 2.35 per cent and 3.13 per cent and that of the Korean firms 2.19 per cent and 2.99 per cent, respectively (OECD, 2006). Thus, the relative R&D intensities of Japanese and Korean firms are quite similar to each other. At the same time, these intensities are the highest among the major OECD countries, clearly exceeding those of the USA and the leading European economies.

Figure 1 shows the composition of corporate R&D in Japan and Korea by industries. In both countries, the relatively highest amount of R&D spending falls to the electronic and microelectronic industries, followed by the automobile industry. It can also be seen, however, that the concentration of R&D in these industries is much stronger in Korea, whereas in Japan a considerable amount falls to other areas, such as general machinery, pharmaceuticals, and chemicals.

Furthermore, some differences between the two countries can also be observed regarding the concentration of R&D expenditures in large firms in general and on the biggest firms in particular. As regards the distribution of R&D expenditures among small and large firms (Figure 2), the majority of corporate R&D is conducted in both countries by relatively large firms with at least 1,000 employees. Moreover, in contrast to widespread perceptions that large firms are particularly dominant in Korea, the relative amount of R&D spending by smaller firms with less than 300 employees is much higher in Korea than in Japan. At the same time, however, the concentration of R&D expenditures on the very largest firms is stronger in Korea, where the top five companies spend more than 40 per cent of the country's total corporate R&D. This is more than in Japan, where the corresponding proportion is less than 20 per cent (see Figure 3). In fact, the R&D expenditures of Samsung Electronics, Korea's biggest R&D spender, amounted to 4.79 billion Won in 2004 alone (Samsung Electronics, 2005), which was equivalent to 28.1 per cent of Korea's total industrial R&D expenditures in this year.

Taken together, the statistical data indicates that whereas the concentration of R&D activities on large firms in general is stronger in Japan, the relative importance of the very largest firms is higher in Korea. Moreover, Korea's corporate R&D is also concentrated to a higher degree than Japan's on a few industries, particularly electronics and microelectronics.

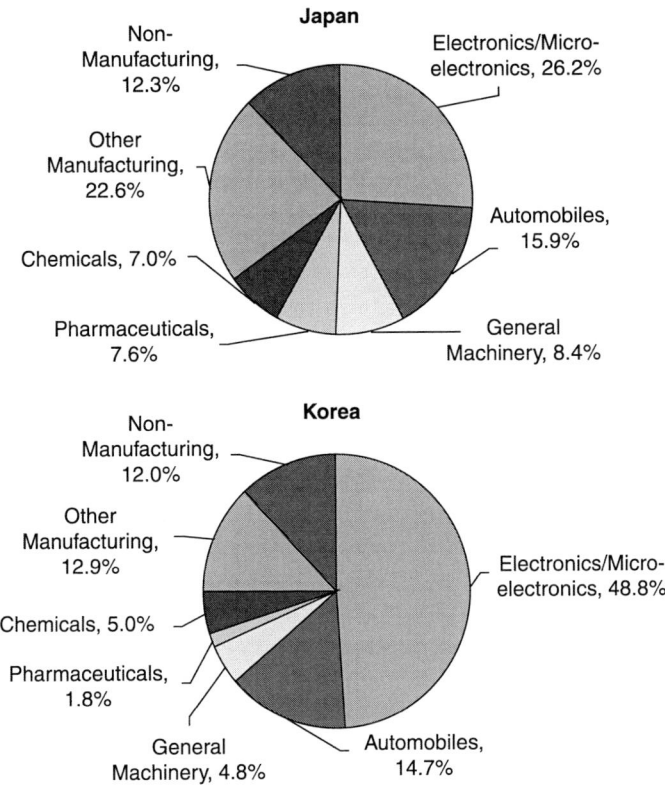

Figure 1. R&D expenditures of Japanese and Korean firms by industry (2004)
Source: Author's composition based on data from MPM (2006); MoST (2005).

In addition to the structure of innovation-related activities conducted by Japanese and Korean firms, various aggregated indicators can also measure their outcomes and performance. In 2006, the US Patent and Trademark Office granted 39,868 patents to Japanese and 6,317 patents to Korean assignees.[1] When set in relation to the two countries' economic size, 9.7 patents were granted per one billion US dollars of purchasing power parity adjusted GDP in the Japanese case and 5.5 patents in the Korean case.[2] This indicates that, notwithstanding Korea's continuous technological catch-up, Japanese firms are on the average still considerably stronger in the production of intellectual property than Korean firms. Furthermore, as regards the two countries' technology balance of payments, Japan's receipts for the export of technology amounted to 2.68 times of its payments for imported technology in 2003, whereas in the case of Korea the receipts were only equivalent to 0.25 times of the payments (OECD 2006). In other words, Japan is a major net exporter and Korea a major net importer of technology. Moreover, the payments for imported technology amounted in 2003 to 3.6 per cent of the business sector's total R&D expenditures in Japan and to 20.2 per cent in Korea (OECD, 2006). This data shows clearly that technology imports still play a very important role for Korean firms, whereas their relative importance is much lower for Japanese firms.

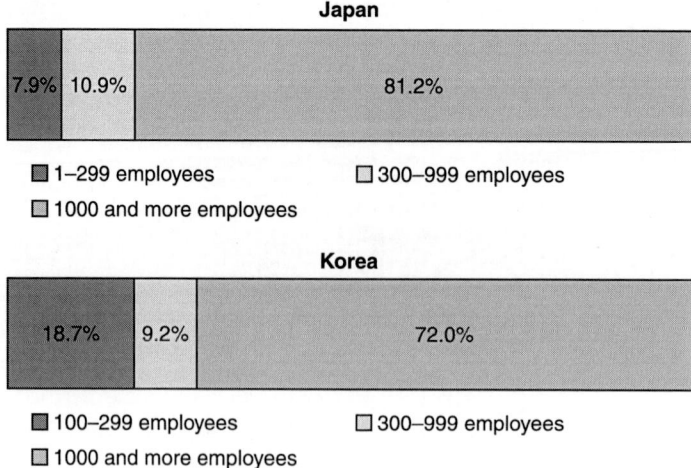

Figure 2. R&D expenditures of Japanese and Korean firms by size (2004)
Source: Author's composition based on data from MPM (2006); MoST (2005).

A closer look at the two countries' competitive positions in high technology industries reveals further information (see Table 1). First, it can be seen that both Japan and Korea hold strong OECD export market shares in the electronics and office/machinery industries and are relatively weak in the pharmaceutical and aerospace industries. Only in the instruments industry, there is a strong difference between the two countries: Japanese firms hold a global export market share of more than 15 per cent, compared with only 2.3 per cent falling to Korean firms. Furthermore, it can be observed that in all high technology industries the Japanese export market share is bigger than the Korean, which is not surprising when considering the much bigger size of the Japanese economy.

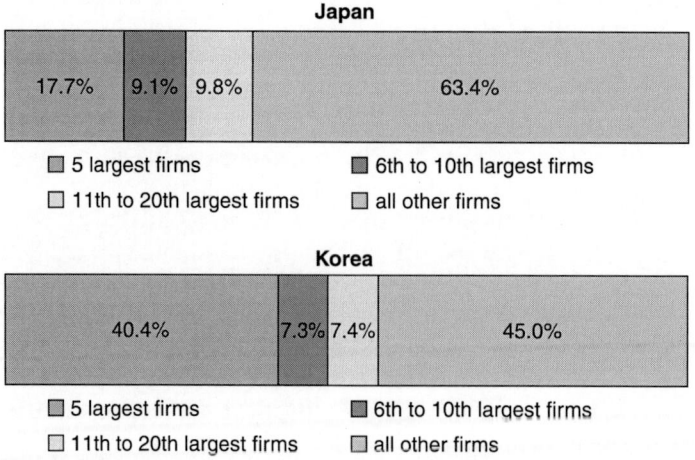

Figure 3. Concentration of R&D expenditures of Japanese and Korean firms (2004)
Source: Author's composition based on data from MPM (2006); MoST (2005).

Table 1. Japan's and Korea's export performance in technology intensive industries (2004)

	Japan		Korea	
	OECD export market share (%)	OECD export market share/ OECD share of country's GDP	OECD export market share (%)	OECD export market share/ OECD share of country's GDP
Electronics	18.4	1.60	13.8	4.48
Office machinery and computers	11.0	0.96	9.2	2.99
Pharmaceuticals	1.9	0.17	0.3	0.10
Instruments	15.1	1.31	2.3	0.75
Aerospace	1.3	0.11	0.3	0.10

Source: Author's composition based on data from OECD (2006).

When dividing the two countries' export market shares by their share of the total GDP among the OECD countries, however (see Table 1), it can be seen that the relative export specialization on the electronics and office machinery/computer industries is much stronger in Korea than in Japan. At the same time, the relative export weakness of the Korean pharmaceutical and aerospace industries is also more pronounced than that of their Japanese counterparts. Taken together, it can be seen that whereas both countries' firms have pronounced, and largely similar, patterns of specialization within the technology intensive sectors, the degree of specialization is even stronger in Korea than in Japan.

In sum, important similarities of Japanese and Korean firms in the field of innovation management have been identified through the review of aggregated statistical data. These are: 1) the R&D intensities of Japanese and Korean firms are similar to each other and at the same time the highest among the major OECD countries; 2) corporate R&D efforts are strongly concentrated on a few industries (electronics, microelectronics and automobiles) in both countries; 3) most R&D is conducted in large firms both in Japan and Korea. At the same time, some major differences have also been observed: 1) the input- and output-related specialization on specific industries is even stronger in Korea than in Japan; and 2) Korean firms possess relatively less intellectual property than Japanese firms and rely to a higher extent on technology imports.

Comparative Analysis by Managerial Field

Strategic Behaviour

The technology strategy of firms can be described in general as their portfolio selection of investments into the development of new products and processes in response to their technological position and the importance of new technologies (Burgelman *et al.*, 2004). Tidd *et al.* (2005) refer in this context specifically to the extent a firm makes entrepreneurial investments into new technologies. In this sense, both Japanese and Korean firms may be characterized as strongly entrepreneurial since their average R&D intensity is higher than that of their rivals from all other leading countries, as has been shown in the previous section.

The perspective can be narrowed down, however, to the extent to which firms are willing to invest into the development of technologies that are new and unfamiliar to them at a given point of time. If they invest strongly into such technologies they are not familiar with, their strategic behaviour can be described as risk taking, whereas it is risk averse, in contrast, if they focus on investments into technologies that are relatively well known to them.

In this regard, some clear differences can be observed between Japanese and Korean firms. Japanese firms have become known as technological forerunners in various fields, such as electronics, microelectronics and automobiles. Notably, however, their R&D efforts have been persistently focused on a given set of technologies in most cases. In the semiconductor industry, for instance, they have stuck mostly to technologies they have been relatively familiar with – both regarding the final products they have been focusing on (mostly memory chips) and supporting process technologies (Okimoto & Nishi, 1994; Shindō, 2006). Likewise, in the automobile industry, the competitive strength of leading firms such as Toyota is mainly built on the perfection of process management practices which were introduced several decades ago rather than being primarily the result of endeavours into new product or process technologies (Fujimoto, 2003). In the pharmaceutical industry, innovations by Japanese firms were also, in contrast to those by firms from other countries, mainly modification-based, that is, built on familiar compounds and applications (Hara, 2005).

Even in fields where Japanese firms have become known as global pioneers, such as liquid crystal displays, their formation of technological knowledge was found to be highly cumulative rather than being the result of large-scale investments into new technologies in a short period of time (Numagami, 1999). Interestingly, studies on corporate strategies of Japanese firms have found that their degree of diversification is relatively low (Kurokawa *et al.*, 2005), thereby giving further support to the observation that they tend to stick to fields of activities they are already familiar with. Taken together, the strategic behaviour of Japanese firms in the field of innovation management appears to be mostly risk averse and conservative.

Studies on the technology strategies of Korean firms suggest that their strategic behaviour is different from that of the Japanese firms. For instance, Hyundai Motor, in its efforts to achieve technological independence from foreign automobile companies, repeatedly developed technologies that it was totally unfamiliar with before, such as building engines (Kim, 1998). Samsung Electronics also has a record of aggressively advancing into new technological fields in the semiconductor industry (Shin & Jang, 2005). Furthermore, the company is actively exploring technological opportunities in fields which are currently still beyond the horizon of commercialization (Cho *et al.*, 2005). Another example is the aggressive, risky and eventually successful development of CDMA cellular phone systems by Korean telecommunication companies (Lee & Lim, 2001). Korean firms have also become known in general for their aggressive diversification strategies on the corporate and business group levels

(Chang, 2003). Their bold advance into new industries, products and technological fields can be characterized as entrepreneurial (Cho *et al.*, 1998). In sum, Korean firms, unlike Japanese firms, tend to adopt a risk taking strategic behaviour in the field of innovation management, as they advance rapidly into technological fields they have not been familiar with hitherto.

Technology Sourcing

The field of technology sourcing is of central importance for the innovation management of firms, as it determines from where and how the technological knowledge that is needed for the development of new products and processes is acquired. In particular, the relative weight of internal and external sources of technology is a crucial issue in this context.

According to the national R&D statistics of the two countries, Japanese and Korean firms spent 13.1 per cent and 13.7 per cent of their total R&D in 2004 externally, respectively (MoST, 2005; MPM, 2006). This indicates that only a relatively minor part of the technology in the corporate sector is sourced from outside. In other words, Japanese and Korean firms appear to acquire most of their new technology through internal activities. This assessment has also been confirmed by innovation surveys in both countries (Gotō & Nagata, 1997; Eom *et al.*, 2005). However, as regards Korean firms, an increase of technological collaboration, particularly with foreign partners, has been observed in recent years (Hobday *et al.*, 2004).

As licensing payments to outsiders are not classified as R&D expenditures, however, the national R&D statistics reveal only part of the overall picture. In the previous overview section it has already been reported that the payments of Korean firms for the import of technology amount to more than one-fifth of their R&D expenditures, whereas the corresponding proportion is very small in the Japanese case.

Taken together, the aggregate data suggests that whereas the firms in both countries give clear overall priority to internal technology sourcing, the Korean firms rely to a much higher degree on external technology sourcing from overseas than the Japanese firms.

These observations can be complemented by survey data collected from corporate R&D managers in both countries. In Figure 4, some results of a questionnaire survey conducted by the author regarding technology sourcing methods of Japanese and Korean firms are summarized. The respondents to this survey were home country based senior R&D managers of leading high technology firms in the semiconductor, telecommunication and pharmaceutical industries in the two countries, with each respondent being responsible for a different technological field. They were requested to rate the relative importance of various technology sourcing methods in their fields of activity on 5-point Likert scales.

The survey results confirm some similarities and differences regarding the technology acquisition of leading firms in the two countries that were identified above. Internal R&D activities were clearly assessed as the most important method of technology sourcing both by the Japanese and the Korean managers. Moreover, methods of internal technology sourcing were in general rated as more

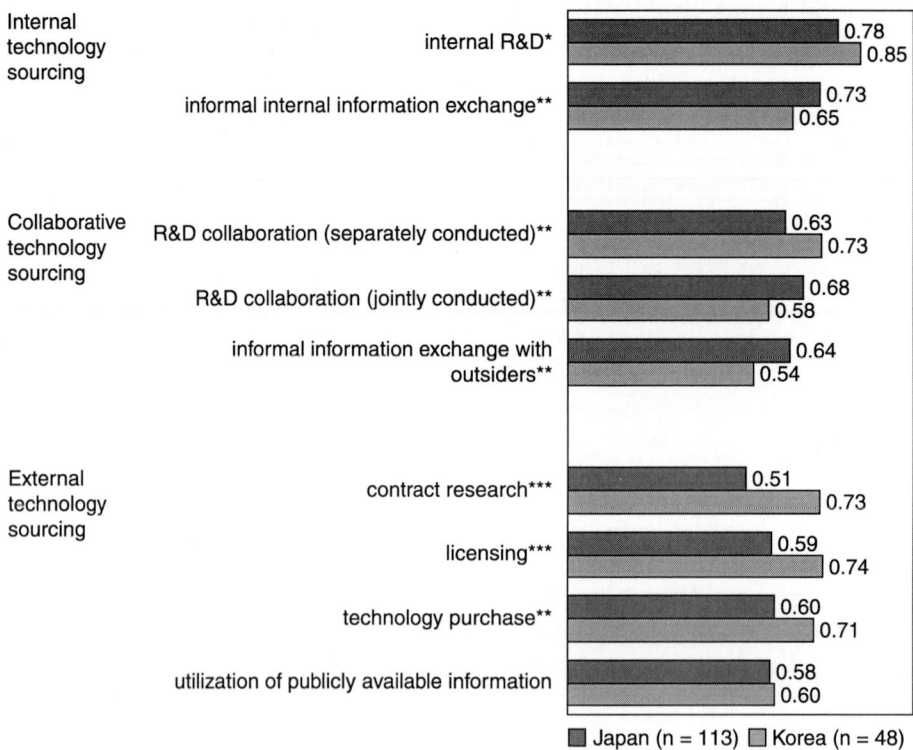

Figure 4. Technology acquisition methods by managers of Japanese and Korean firms (mean values on a standardized scale between 0 and 1) * = significant difference at 0.1 level; ** = significant difference at 0.05 level; and *** = significant difference at 0.01 level

important than collaborative technology sourcing and external technology sourcing in both countries. At least as regards Japanese firms, this tendency to rely to a relatively high extent on internal R&D when developing new technologies has also been confirmed by other empirical research. For example, Japanese firms were found to be engaged in far fewer technology alliances than firms from other leading countries (Sakakibara, 2005).

Alongside with these similarities, however, two differences can also be observed. First, the Korean managers perceive informal information exchange (both internally and externally) as methods of technology sourcing clearly as less important than the Japanese managers, suggesting that they rely to a relatively higher extent on formal projects as regards the acquisition of new technology. Second, various external technology sourcing methods, namely contract research, licensing and technology purchasing were perceived as much more important by the Korean respondents than by the Japanese respondents, thereby confirming the author's previous observations derived from international technology trade data of the two countries, as well as results of case studies on Korean firms (Hobday *et al.*, 2004).

Another important aspect of international technology sourcing is the amount and quality of technological knowledge acquired from overseas sources. Aside from data on international technology trade, this dimension can also be measured by the relative importance of overseas R&D expenditures among the total R&D expenditures of firms. According to the OECD (2006), Japanese firms spent 1.8 per cent and 4.3 per cent of their total R&D abroad in 1998 and 2003, respectively. Although the importance of international R&D has thus been increasing for Japanese firms in recent years, its relative weight is still much lower than in the case of US firms, which spend about 15 per cent of their R&D abroad, and of firms from the main European countries, where more than 20 per cent of total corporate R&D is spent overseas.

As regards Korean firms, there is no such comprehensive data on international R&D expenditures available. A recent survey among 192 leading Korean firms revealed, however, that only 15.6 per cent of these firms maintained any overseas R&D activities (J. Lee *et al.*, 2005). Moreover, more than half of those firms that were active in international R&D spent less than five per cent of their total R&D abroad. These results, although they are not directly comparable with the OECD data on Japan and other countries, indicate that the overall importance of international R&D is also quite low for the Korean firms.

The strategic direction of international R&D activities also needs to be considered in the context of technology sourcing, however. Not all international R&D is primarily directed at sourcing technology from abroad. In this regard, a distinction can be made between home-base augmenting and home-base exploiting international R&D (Kuemmerle, 1997). The main intent of the former is sourcing technology from abroad and transferring it to the home country, whereas the latter is primarily directed at the application and exploitation of technology acquired at home on overseas markets. In the survey of international R&D by Korean firms mentioned above, the by far most frequently cited reason for conducting international R&D was the 'development of cutting-edge technology through the acquisition of advanced technology' (J. Lee *et al.*, 2005: 69). This suggests that the majority of overseas R&D conducted by Korean firms is targeted at technology sourcing from abroad. Qualitative studies on the international R&D of Japanese firms indicate, in contrast, that in many cases the related activities appear to be home-base exploiting rather than home-base augmenting (Gerybadze & Reger, 1999).

Taken together, the review of survey data on the technology sourcing of Japanese and Korean firms indicate that 1) firms from both countries give the highest priority to internal technology sourcing; 2) informal technology sourcing methods are more important for Japanese than for Korean firms; 3) the relative importance of external technology sourcing from abroad is much higher for Korean than for Japanese firms; and 4) whereas the weight of international R&D is relatively low among firms from both countries, the Korean firms' activities in this field appear to be more strongly oriented towards international technology sourcing than those of the Japanese firms.

R&D Management Practices

There are various managerial practices that appear to be widely implemented in the field of R&D both by Japanese and Korean firms. Firms from both countries have become well known for their fast development of new products that have greatly contributed to their international competitiveness (Fujimoto, 2003; Shin & Jang, 2005). The shortening of product development time was achieved through parallel and overlapping work at different development stages instead of sequential work organization (Cho *et al.*, 1998; Yasumoto & Fujimoto, 2005; Nobeoka, 2006) and the integration of development work across different functions and departments within a firm through the creation of cross-functional teams and of special task forces (Yasumoto & Fujimoto, 2005; Shin & Jang, 2005). Accordingly, product managers who coordinate such cross-functional teams or task forces appear to have a strong position within the firms' management systems when compared with functional department managers (Fujimoto, 2003). Thus, it can be concluded that Japanese and Korean firms share many practices and tools in R&D management that seem to have contributed greatly to their competitive strength during the last decades. At the same time, however, some differences can also be observed between them.

In the case of Japanese firms, the early and intensive involvement of independent core suppliers, that is, such firms that may have an intimate relationship with certain producers but are not managerially controlled by them, into development activities is a another widespread managerial practice (Numagami, 1999; Fujimoto, 2003). Korean firms are less known for this approach. In their case, the integration of suppliers into development projects appears to be limited mainly to such firms that are members of the same business group and are thereby controlled managerially (Chang, 2003).

Some Korean firms have become known recently, however, for their intensive efforts to improve the marketability of their new products through the early integration of design activities in product development projects (Cho *et al.*, 2005). In various corporate design centres which are often located overseas, possible future product designs are created at the very beginning of the whole new product development process. This 'design-first' approach has not been widely reported for Japanese firms.

Taken together, Japanese and Korean firms share various R&D management practices such as parallel development processes and cross-functional integration that have enhanced their technological competitiveness to a great extent. At the same time, however, some differences between them can also be observed. Specifically, Japanese firms appear to place a particular emphasis on the upstream integration in development projects through the early involvement of suppliers, whereas Korean firms put a particular focus on downstream integration by positioning new product designs as the starting points of new product development projects.

Human Resource Management Practices

Finally, human resource management practices also play a central role in innovation management, as human skills and efforts constitute the backbone of innovations. Traditionally, long-term employment and seniority based promotion and compensation appears to have been dominant practices of human resource management both in Japanese and Korean firms, particularly large companies. Employment relations with managers and workers were stable, and promotions and pay increases were predominantly based on the number of years an individual has worked for one particular company rather than on his or her individual performance (Westney, 1994; Chung *et al.*, 1997).

Japanese firms have started to experiment with partially different employment practices for their R&D staff since the 1990s. Specifically, the introduction of partially performance-based compensation systems and of result-oriented assignments has been taken into consideration (KSKKK, 1994). However, these changes in the employment and compensation systems of firms have been implemented only very gradually and cautiously so far. For example, only 13.9 per cent of all firms who responded to a survey by the Japanese Ministry of Health and Labor have introduced performance-based annual compensation systems until 2005 (Kōsei Rōdōshō, 2005). Moreover, it has been observed that even those companies which introduced such performance-based systems in the area of R&D typically implemented the new schemes only on a limited scale (for example, for a few dozen senior researchers in large firms) rather than for their R&D staff in general (KSKKK, 1994). Taken together, it appears that whereas some Japanese firms have started to change their human resource management systems – particularly in the field of R&D–from the traditional seniority orientation towards meritocracy orientation, this shift can be characterized as a slow and cautious one (Benson & Debroux, 2003).

In contrast, the Korean case demonstrates how firms have changed their human resource management practices (which were quite similar to those of the Japanese firms until the 1990s) more rapidly in recent years. This change was triggered by the Asian financial crisis of 1997 after which many large Korean companies and business groups had to lay off managers and workers on a large scale or went bankrupt altogether (Park & Yu, 2002). Thus, the stability of employment relations that existed in most large Korean firms during the previous decades was disrupted as a result of an external economic shock.

However, many Korean firms changed their human resource management systems also after recovery from the crisis. According to the Korean Ministry of Labor, 48.4 per cent of all surveyed enterprises have introduced performance-based annual compensation systems and 32.1 per cent of them had set up profit-sharing systems by 2005, whereas only 23 per cent and 20.6 per cent of them had such systems in 2000, respectively (Nodongbu, 2005). This clearly indicates a widespread departure from the previous seniority-based compensation and promotion system. In addition, some large firms have also adopted internal corporate venturing programmes (Bae & Rowley, 2003) and determine the compensation employees receive through consideration of their individual and their organizational unit's performance (Cho *et al.*, 2005). Likewise, 'high risk, high return' incentive structures that seek to motivate employees through financial incentives, such as stock options, have also been widely adopted by Korean venture firms (Bae & Yu, 2005). Taken together, although traditional, seniority based compensation and promotion systems are still existent in Korean firms, a transition towards stronger performance orientation in human resource management, particularly in the field of R&D, has clearly been faster than among Japanese firms.

Discussion

The main findings of the previous comparative analysis regarding the innovation management of Japanese and Korean firms by managerial field are summarized in Table 2. It can be seen that whereas certain managerial similarities among firms from the two countries exist, there are some remarkable differences that set them apart.

A number of historical, cultural and institutional factors contribute to explain the commonalities and the differences between firms from the two countries. The Japanese and Korean firms' propensity to make relatively high investments into R&D and to give priority to internal technology sourcing can be associated with the strong strategic intent that has been widely observed among firms from East Asian countries in the areas of strategic management and innovation management (Hamel & Pralahad, 1989). Firms from this region have become known for their strong cohesiveness among managers and workers that facilitated the successful implementation of aggressive strategies to catch up with and overtake their main competitors, often on technological grounds.

One underlying cultural feature for this orientation is the collectivism in East Asia. In collectivistic societies, the distinction between group members and outsiders is more important and relationship building and social interaction among group members is given higher priority over interaction with outsiders than in individualistic societies (Triandis, 1995). As both Japanese and Korean societies are regarded as collectivistic (Cha, 1994; Yamaguchi, 1994), it is relatively easy for firms from these countries to develop a strong coherence among their managers and employees which may result in a tendency to prefer technology sourcing through internal R&D projects over the sourcing of technological knowledge from outside. Moreover, another common cultural feature which also helps to explain the high R&D investments by firms from both countries is the strong 'long term orientation' by Japanese and Koreans (Hofstede, 2001), as R&D activities can be classified as investments into the future, and it often takes a long time to receive a return on them.

The commonalities among Japanese and Korean firms in R&D management practices and their traditional preference for long-term employment and seniority-based compensation and promotion can also be linked to similar underlying factors. The strong internal coherence of Japanese and Korean companies as supported by cultural factors helps them to establish managerial practices that rely on intensive internal (including cross-functional) communication and coordination. Long-term employment and seniority based compensation and promotion practices can be rooted to skill formation systems which are primarily company oriented rather than profession oriented (Aoki, 1988), as well as to East Asian cultural traditions such as Confucianism which place high value on seniority (Chung et al., 1997; House et al., 2004). Moreover, as Korea was occupied by Japan for 35 years in the first half of the twentieth century, Korean management systems have been influenced by Japanese management practices, particularly in areas such as human resource management where both countries share common cultural roots (Chen, 2004).

Similarly, contextual factors also help to explain the differences in innovation management between Japanese and Korean firms. As regards the more risk taking strategic orientation of Korean firms as compared to Japanese firms, differences

Table 2. Features of innovation management by Japanese and Korean firms

Managerial field	Japanese firms	Korean firms
Strategic behaviour	High investment in R&D Risk-averse strategic orientation	High investment in R&D Risk-taking strategic orientation
Technology sourcing	High priority for internal R&D High importance of informal technology sourcing methods Low degree of international technology sourcing through R&D	High priority for internal R&D Strong reliance on external sources of technology from overseas Modest degree of international technology sourcing through R&D
R&D management practices	Phase-overlapping parallel project management Cross-functional integration Early and strong integration of external suppliers in R&D	Phase-overlapping parallel project management Cross-functional integration Early integration of design in R&D
Human resource management practices	Traditionally long-term employment and seniority-based compensation & promotion Slow transition towards performance-based incentives and compensation	Traditionally long-term employment and seniority-based compensation & promotion More flexible employment since Asian financial crisis Rapid transition towards performance-based incentives and compensation

between the two countries in corporate cultures and corporate governance have to be considered. The management of Japanese firms has been described as strongly consensus driven. Decisions are made only after extensive vertical and horizontal communication processes that constitute the base for an early consensus of all relevant organizational sub-units and individuals (Nonaka & Takeuchi, 1995). Under such circumstances, aggressive, risk taking strategies are difficult to implement because it is likely that at least some relevant actors inside the company do not support them. In addition, an emphasis on long-standing harmonious relationships with suppliers and other associated firms also makes radical changes difficult to implement (Collison & Wilson, 2006).

The Korean management style, in contrast, has been described as strongly hierarchical in general (Chung et al., 1997; Chen, 2004). Top managers can make decisions without needing to seek a broad consensus in the whole organization. The owners of Korean business groups have become particularly well known for having almost unlimited power regarding the strategy formulation and implementation of their group firms (Chang, 2003). As a result, risk taking strategies are more likely to take place because the approval of a larger number of organizational units or individuals is not needed.

The relatively higher reliance of Korean firms on overseas-produced external technology sources can be explained by their status as relative latecomers. Whereas Japanese firms are also regarded as latecomers in their technological skill formation relative to firms from some leading Western countries, they had relatively more time for their technological catch-up than Korean firms. In the Japanese case, this catch-up began essentially in the late nineteenth century and continued well into the second half of the twentieth century (Francks, 1992). In Korea, in contrast, a similar process could only be initiated after retaining national independence in 1945 and, on a more permanent base, after the end of the Korean War in 1953 (Kim, 1997).

As a result, Japanese firms typically attempted to close the technological gap with the leading countries by fully absorbing and internalizing not only some specific core technologies, but also a wide range of supporting technologies, such as methods to produce parts and components needed for the manufacturing of complex final products in order to attain full technological independence from other countries. Korean firms, in contrast, provided with less time to catch up technologically, rather chose to focus their internal efforts on a limited number of core technologies and to rely on foreign companies regarding supporting technologies through outsourcing and licensing agreements.

In addition, an economic country size argument can also be applied in this context. As the Japanese economy is one of the biggest in the world, it offers sufficient space to accommodate a broad set of industries that can compete both in a large domestic market and internationally. The Korean economy is considerably smaller,[3] thereby forcing companies to concentrate on a narrower range of industries and technological fields in which the country can compete internationally. This stronger technological specialization of the Korean economy can be observed regarding two dimensions: horizontally, meaning that it is concentrated on fewer industries, as shown in Figure 1 and in Table 1, and vertically, indicating a specialization on specific steps in the value chain *within* industries. The latter point is visible in the stronger reliance of Korean firms on external knowledge from overseas that is often related to technological fields that are supporting their own core technologies.

The same arguments on Korea's relative latecomer position and its relatively smaller economy can be applied to explain the stronger integration of external suppliers in firms' R&D processes in Japan than in Korea. Due to the later and faster development and the country's smaller economic size, technologically strong supplier industries could not grow up to the same extent in Korea as in Japan. As a result, domestic suppliers tend to be technologically less attractive partners in the Korean case.

The strong design orientation of Korean firms in their innovation management cannot be linked as easily to historical or institutional factors as most of the other results of the previous comparative analysis. It can be observed quite broadly, however, that Koreans have a strong sense for design and fashion. Therefore, this managerial feature of Korean firms also seems to be embedded in a national cultural background.

Finally, the more rapid transition of human resource management from stability and seniority orientation to flexibility and performance orientation in Korea can be partially attributed to the immediate necessity to cope with the economic effects of the Asian financial crisis. Many Korean firms needed to reduce their labour cost and to improve their productivity in the years after the crisis to survive. However, as mentioned in the previous section, the transition of human resource management practices also continued in Korea after the economic aftermath of the crisis has been overcome. A possible reason for this phenomenon lies again in the relatively smaller size of the Korean economy. Since a smaller economy like Korea is in general more dependent on overseas markets and more vulnerable to external shocks than a bigger one like Japan, the managers of Korean firms may have felt a stronger sense of urgency to change their managerial practices in order to restore international competitiveness.

Managerial Evaluation

How can the features of Japanese and Korean innovation management that were identified above be evaluated from a managerial perspective? Managerial strategies and measures are generally implemented in response to the specific competitive and financial situation of a firm. Therefore, a managerial practice cannot be classified as universally 'good' or 'bad'. Its effectiveness and outcomes always depend on the concrete case and the situation in which a firm is operating. Notwithstanding this fundamental fact, however, managerial tools and systems can still be evaluated as being successful or unsuccessful in a wide range of industries or competitive contexts.

Some of the common managerial tools and practices adopted by Japanese and Korean firms have apparently contributed greatly to their innovativeness and competitive success over the last few years and decades. Strong investments into R&D and giving high priority to internal technology sourcing were strategies which fitted well into an East Asian cultural environment which supports the formation of organizations with a strong strategic intent to catch up with and overtake competitors technologically. Likewise, on the implementation level, practices like phase-overlapping project management and cross-functional integration also proved to be very successful in a wide range of industries, such as automobiles, electronics and microelectronics (Nonaka & Takeuchi, 1995; Kim, 1998; Shin & Jang, 2005; Nakata *et al.*, 2006). Moreover, long-term employment practices, combined with seniority based compensation and promotion systems are also likely to have enhanced the technological performance of Japanese and Korean firms in a catchup context, as they provided job security to managers and workers and increased their working motivation.

In recent years, however, these human resource management practices have been evaluated more critically, particularly in the context of innovation management. Whereas they may have been supportive during the catch-up period, some disadvantages became apparent once the firms reached the technological forefront. Under such circumstances, the individual creativity and performance of researchers and engineers gains in importance compared with an organization's overall coherence. Moreover, flexible employment practices, such as hiring and compensating capable researchers from outside on a competitive base regardless of their age, have the advantage of bringing diversity and new ideas and perspectives into an R&D organization. Taken together, it seems that the environment and the challenges for the human resource management of Japanese and Korean firms, particularly in innovation related areas, have changed considerably due to their successful technological catch-up.

From this perspective, the more rapid transition toward flexible and performance based approaches in human resource management adopted by Korean firms appears to be more promising and effective than the cautious approach chosen by most Japanese firms. In other words, the Korean firms adapt to the new managerial challenges more quickly than the Japanese firms.

The higher reliance of Korean firms on external technology from overseas when compared with Japanese firms appears to be a relative disadvantage at first glance when considering accumulated internal technology as a source of competitive strength. In recent years, however, there has been a general tendency towards more technology outsourcing and collaboration in R&D (Hicks *et al.*, 1996; Tapon & Thong, 1999; Hagedoorn, 2002), indicating that many firms do not perceive internal technology sourcing any more as their first strategic choice regarding all technological fields they need to cover. In certain cases, technology outsourcing may be simply more efficient or effective than internal R&D (Pisano, 1990). Moreover, from a strategic viewpoint, concentrating internal R&D resources on a limited number of core areas can be more advantageous than spreading them broadly on a large number of fields (Roberts & Berry, 1985). Thus, whereas Korean firms appear to be potentially more vulnerable to the behaviour of external partners than Japanese firms due to their higher reliance on external technology from abroad, it cannot be said that this situation puts them at a general disadvantage. Rather, the concrete evaluation depends on the reliability and the quality of the overseas sources of technology a firm is relying on. In fact, some recent surveys strongly indicate that technology outsourcing and external collaboration enhance the innovative performance of Korean firms (Jeong, 2004; Sung, 2005).

As regards the more risk taking strategic behaviour of Korean firms, their aggressive approach has apparently worked very well in a number of cases, as some of them have moved from technologically trailing positions towards the global forefront or even global leadership (Kim, 1998; Shin & Jang, 2005). However, in other cases, the aggressive entry of Korean firms into new technological fields has resulted in failure, as they could not acquire the technological competencies they needed to become competitive in these areas (Lim, 2006). In other words, such risk taking technology strategies are not always successful. The initial resource position of the firm as well as the technological and the market regimes strongly determine the chances for their successful implementation. Moreover, it should also be considered that the risk taking strategic behaviour of Korean firms evolved mostly from technological backward positions hitherto. Whereas the strategic choices of the firms can still be described as risk taking in such cases, as the technological fields they advanced into were new from their own perspective, it can also be argued that such strategies are relatively easy to implement in a catchup process, as the goals are clearly visible and there is less technological and market uncertainty than for a firm in a leading position. Thus, it remains to be seen to what extent Korean firms continue to pursue aggressive technology strategies from leading positions in the future, and how successful these strategies will be under the changed environmental conditions.

Finally, the strong integration of external suppliers in the R&D process in Japan and the early adoption of new designs when developing new products in Korea are regarded as sources of technological and competitive strength in Japanese and Korean firms, respectively (Clark & Fujimoto, 1991; Cho *et al.*, 2005). In fact, it appears that in both cases, country-specific features have resulted in favourable conditions for firms that have their home base in each country. The technologically strong supplier base is the result of the industrial development of Japan in the last 50 years, as large Japanese firms have preferred to upgrade the technological competencies of their suppliers through technological transfer and close collaborative relationships with them instead of integrating backwards (Fujimoto, 2003). The early integration of design-related knowledge in development processes by Korean firms is possibly the result of the generally strong sense Koreans have regarding design and fashion related developments that may have evolved from various cultural and climatic features, though it seems that this aspect has not been addressed yet by cross-cultural research or by the management literature in general. Both in Japan and Korea, however, the backgrounds of the observed managerial features are embedded in the economic systems of the respective countries and thus appear to constitute cases of 'national competitive advantage' (Porter, 1990).

Implications for Comparative Management Research and Practice

The analysis in this study has various limitations. First, due to the broad scope of the topic that was covered, most of the findings are based on the review of secondary data sources and rigorous quantitative research methods could not be applied. Second, many of the observations are derived from a relatively limited number of leading Japanese and Korean firms. Therefore, any generalizations of the results should be made with caution. Third, regarding some of the managerial areas that have been reviewed, there is in general a scarcity of empirical data. Taken together, the findings presented in this study should be regarded as tentative ones that need to be validated by further research.

There are several promising directions for such further comparative analysis of Japanese and Korean firms. In-depth, contrasting case studies of a few leading firms from both countries could enhance our understanding about similarities and differences between their management styles. Moreover, studies which focus on specific industries where both countries play a leading role (for example, in business sectors such as electronics, microelectronics or automobiles) could constitute further valuable contributions to this field of research. As the existing research is mostly concentrated on large firms, comparative studies on the innovation management of small and medium-sized firms and venture firms in Japan and Korea would also advance our knowledge regarding this field. Finally, comparative studies which embrace not only Japan and Korea, but also firms from other leading East Asian economies such as Taiwan, Hong Kong and Singapore could broaden the perspective further and address the question of whether there are any commonalities in the management styles of East Asian firms in general which set them apart from their rivals in western countries.

From a managerial perspective, some implications for Japanese firms, Korean firms and firms from other Asia-Pacific countries can be derived from this study. Managers of Japanese firms should consider speeding up the transition of their human resource management systems in innovation related activities towards more flexibility and performance orientation, as they may suffer competitive disadvantages in the future if they are left behind in this field by dynamic competitors from countries like Korea. Moreover, given the high level of internal technological competencies many Japanese firms have accumulated, their management may reconsider its strategic behaviour in certain cases. Whereas strategic conservatism appears to be a safe approach at a first glance, firms that stick to this behaviour may become gradually marginalized by their competitors in dynamic industries, as they may miss valuable opportunities for future growth. One such case appears to be the semiconductor industry where Japanese firms have continuously lost international market share to US and Korean firms during the last 15 years, as explained by Okada in this current collection.

From the viewpoint of Korean firms, the findings of our analysis suggest that whereas pursuing aggressive technology strategies may often be a promising approach, it does not improve a firm's competitive position unconditionally. When the market and technological regimes and a firm's internal capabilities are at odds against such strategies, they are almost certainly doomed to failure. Thus, managers of Korean firms should consider these internal and external factors before deciding to advance into new, unfamiliar technological fields.

Finally, managers of firms from other, particularly western, countries should be aware of the commonalities and differences between Japanese and Korean firms when dealing with them as competitors or partners in strategic alliances. Japanese and Korean firms are both very advanced and effective regarding the implementation of various R&D management practices. However, considerable differences set them apart in areas like strategic behaviour and technology sourcing. Therefore, their innovation management should never be assumed to be fundamentally similar in general.

Acknowledgements

This research was supported by a Korea University grant. The author would like to thank the *APBR* special issue editors and reviewers for their helpful comments on earlier versions of this study.

Notes

[1] The patent information has been retrieved from the online server of the US Patent and Trademark Office, available at www.uspto.gov/patft/index.html.

[2] Author's calculation based on the patent data from the US Patent Trademark Office and Purchasing Power Parity adjusted GDP data from OECD (2006).

[3] According to the World Bank's statistics, the Korean GDP amounted in 2005 to 17.5 per cent and 26.8 per cent of the Japanese GDP on a nominal base and a purchasing power parity adjusted base, respectively (IBRD, 2006).

References

Aoki, M. (1988) *Information, Incentives, and Bargaining in the Japanese Economy* (Cambridge: Cambridge University Press).

Bae, J. & Rowley, C. (2003) Changes and continuities in South Korean HRM, *Asia Pacific Business Review*, 9(4), pp. 76–105.

Bae, J. & Yu, G.-C. (2005) HRM configurations in Korean venture firms: resource availability, institutional force and strategic choice perspectives, *International Journal of Human Resource Management*, 16(9), pp. 1759–1782.

Benson, J. & Debroux, P. (2003) Flexible labour markets and individualized employment: the beginnings of a new Japanese HRM system, *Asia Pacific Business Review*, 9(4), pp. 55–75.

Burgelman, R. A., Christensen, C. M. & Wheelwright, S. C. (2004) *Strategic Management of Technology and Innovation*, 4th ed. (Boston, MA: McGraw-Hill).

Cha, J.-H. (1994) Aspects of individualism and collectivism in Korea, in: U. Kim, H. C. Triandis, Ç. Kâğitcibaşi, S.-C. Choi & G. Yoon (Eds) *Individualism and Collectivism: Theory, Methods, and Applications*, pp. 157 174 (Thousand Oaks: Sage).

Chang, S.-J. (2003) *Financial Crisis and Transformation of Korean Business Groups, The Rise and Fall of Chaebols* (Cambridge: Cambridge University Press).

Chen, M. (2004) *Asian Management Systems: Chinese, Japanese and Korean Styles of Business*, 2nd ed. (London: Thomson).

Cho, D.-S., Kim, D.-J. & Rhee, D. K. (1998) Latecomer strategies: evidence from the semiconductor industry in Japan and Korea, *Organization Science*, 9(4), pp. 489–505.

Cho, H., Chun, H. & Lim, S. (2005) *Dijital chungbokcha Samsung Chuncha* (Digital Conquerer Samsung Electronics) (Seoul: Maeil Kyungjae Sinmunsa).

Chung, K. H., Lee, H. C. & Jung, K. H. (1997) *Korean Management: Global Strategy and Cultural Transformation* (Berlin: de Gruyter).

Clark, K. B. & Fujimoto, T. (1991) *Product Development Performance: Strategy, Organization, and Management in the World Auto Industry* (Boston, MA: Harvard Business School Press).

Collison, S. & Wilson, D. C. (2006) Inertia in Japanese organizations: knowledge management routines and failure to innovate, *Organization Studies*, 27(9), pp. 1359–1387.

Eom, M., Choi, C. & Lee, C. (2005) *2005-nyeondo hanguk oi gisul hyeoksin chosa: checho-ob bumun*, (2005 Korean Innovation Survey: The Manufacturing Sector) Chosa Yongu 2005–05 (Survey Research No. 5, 2005) (Seoul: STEPI).

Francks, P. (1992) *Japanese Economic Development: Theory and Practice* (London: Routledge).

Fujimoto, T. (2003) *Nōryoku kōchiku kyōsō* (Competition Based on Construction of Capabilities) (Tokyo: Chūō Kōron Shinsha).

Gerybadze, A. & Reger, G. (1999) Globalization of R&D: recent changes in the management of innovation in transnational corporations, *Research Policy*, 28(2/3), pp. 251–274.

Gotō, A. & Nagata, A. (1997) *Inobēshon no sen'yū kanōsei to gijutsu kikai: Sābei dēta ni yoru nichi-bei hikaku kenkyū* (Technological Opportunities and Appropriating the Returns from Innovation: Comparison of Survey Results from Japan and the U.S.), NISTEP Report No. 48 (Tokyo: Kagaku Gijutsu Seisaku Kenkyūsho).

Hagedoorn, J. (2002) Inter-firm R&D partnerships: an overview of major trends and patterns since 1960, *Research Policy*, 31(4), pp. 477–492.

Hamel, G. & Pralahad, C. K. (1989) Strategic intent, *Harvard Business Review*, 67(3), pp. 63–76.

Hara, T. (2005) Innovation management of Japanese pharmaceutical companies: the case of an antibiotic developed by Takeda, *International Journal of Technology Management*, 30(3/4), pp. 351–364.

Hicks, D. M., Isard, P. A. & Martin, B. R. (1996) A morphology of Japanese and European corporate research networks, *Research Policy*, 25(3), pp. 359–378.

Hobday, M., Rush, H. & Bessant, J. (2004) Approaching the innovation frontier in Korea: the transition phase to leadership, *Research Policy*, 33(10), pp. 1433–1457.

Hofstede, G. (2001) *Culture's Consequences*, 2nd ed. (Newbury Park, NJ: Sage).

House, R. J., Hanges, P. J., Javidan, M., Dorfman, P. W. & Gupta, V. (2004) *Culture, Leadership, and Organizations: The GLOBE Study of 62 Societies* (Thousand Oaks, CA: Sage).

IBRD (International Bank for Reconstruction and Development) (2006) *World Development Indicators 06* (Washington, DC: The World Bank).

Jeong, J. (2004) Gisul doibi giob gachie michinun yeonghyang (The effect of technology introduction on the firm value), *Gisul Hyeoksin Yongu*, 12(1), pp. 49–65.

Kim, L. (1997) *Imitation to Innovation: The Dynamics of Korea's Technological Learning* (Boston, MA: Harvard Business School Press).

Kim, L. (1998) Crisis construction and organizational learning: capability building in catching-up at Hyundai Motor, *Organization Science*, 9(4), pp. 506–521.

Kōsei Rōdōshō (2005) *Heisei 17-nen shurō jōken sōgō chōsa* (Overall investigation of employment conditions 2005) (Tokyo: Kōsei Rōdōshō).

KSKKK (Kikai Shinkō Kyōkai Keizei Kenkyūjō) (1994) *Minkan kigyō no kenkyŪ kaihatsu katsudō ni kan suru kiso chōsa* (Basic Investigation of the R&D Activities of Private Enterprises) (Tokyo: Kikai Shinkō Kyōkai Keizei Kenkyūjō).

Kuemmerle, W. (1997) Building effective R&D capabilities abroad, *Harvard Business Review*, 75(2), pp. 61–70.

Kurokawa, S., Pelc, K. I. & Fujisue, K. (2005) Strategic management of technology in Japanese firms: literature review, *International Journal of Technology Management*, 30(3/4), pp. 223–247.

Lee, J. W., Lee, J.-O. & Kim, K. K. (2005) *R&D gurobolwha: hyeonghwang gwa suchun jukcheong ul uihan chipyo kebal* (Development of R&D Globalization Indicators) Cheongjek Yongu 2005–09 (Policy Research no. 9, 2005) (Seoul: STEPI).

Lee, K. & Lim, C. (2001) Technological regimes, catching-up and leapfrogging: findings from the Korean industries, *Research Policy*, 30(3), pp. 459–483.

Lee, K., Lim, C. & Song, W. (2005) Emerging digital technology as a window of opportunity and technological leapfrogging: catch-up in digital TV by the Korean firms, *International Journal of Technology Management*, 29(1/2), pp. 40–63.

Lim, C. (2006) The difficult catch up in the numerical controller sector, in: M. Hemmert (Ed.) *Emerging Economies in Asia and Europe: New Challenges for Competition and Collaboration, Proceedings of the 23rd Annual Conference of the Euro-Asia Management Studies Association*, pp. 137–156 (Seoul: Korea University Business School).

MoST (Ministry of Science and Technology, Republic of Korea) (2005) *Report on the Survey of Research and Development in Science and Technology, 2005 Edition* (Seoul: MoST).

MPM (Statistics Bureau, Ministry of Public Management, Home Affairs, Posts and Telecommunications Japan) (2006) *Report on the Survey of Research and Development 2005* (Tokyo: Japan Statistical Association).

Nakata, C., Im, S., Park, H. & Ha, Y.-W. (2006) Antecedents and consequence of Korean and Japanese new product advantage, *Journal of Business Research*, 59(1), pp. 28–36.

Nobeoka, K. (2006) Maruchi purojekuto senryaku: Jidōsha no seihin kaihatsu ni okeru purattofōmu manejimento (Multi project strategy: multi platform management in the automobile industry), in: H. Itami, T. Fujimoto,

T. Okazaki, H. Itoh & T. Numagami (Eds) *Rīdingusu nihon no kigyō shisutemu, dai 2-ki, dai 3-maki: Senryakyu to inobēshon* (Readings on the Japanese Firm as a System, II, Vol. 3: Strategy and Innovation), pp. 127–151 (Tokyo: Yūhikaku).

Nodongbu (2005) *Yeonbongjae, seonggwa bebunjae siltae chosa gyolgwa 2005–12* (Survey Results Regarding Annual Compensation Systems and Profit Sharing Systems, December 2005) (Seoul: Nodongbu).

Nonaka, I. & Takeuchi, H. (1995) *The Knowledge-Creating Company* (New York: Oxford University Press).

Numagami, T. (1999) *Ekishō disupurei no gijutsu kakushinshi* (History of liquid crystal display technology) (Tokyo: Hakutō Shobō).

OECD (Organization for Economic Cooperation and Development) (2006) *Main Science and Technology Indicators, Volume 2006/2* (Paris: OECD).

Okimoto, D. I. & Nishi, Y. (1994) R&D organization in Japanese and American semiconductor firms, in: M. Aoki & R. Dore (Eds) *The Japanese Firm: Sources of Competitive Strength*, pp. 178–208 (Oxford: Oxford University Press).

Park, W.-S. & Yu, G.-C. (2002) HRM in Korea: transformation and new patterns, in: Z. Rhee & E. Chang (Eds) *Korean Business and Management: The Reality and Vision*, pp. 367–391 (Elizabeth, NJ: Hollym).

Pisano, G. P. (1990) The R&D boundaries of the firm: an empirical analysis, *Administrative Science Quarterly*, 35(1), pp. 153–176.

Porter, M. (1990) *The Competitive Advantage of Nations* (New York: Free Press).

Porter, M., Takeuchi, H. & Sakakibara, M. (2000) *Can Japan Compete?* (Houndmills: Macmillan).

Roberts, E. B. & Berry, C. A. (1985) Entering new businesses: selecting strategies for success, *Sloan Management Review*, 26(3), pp. 3–17.

Sakakibara, K. (2005) *Inobēshon no shŪekika: Gijutsu keiei no kadai to bunseki* (Profiting from Technological Innovations: Challenges and Analysis of Technology Management) (Tokyo: Yūhikaku).

Samsung Electronics (2005) *Jae 36-ki saob bogoso*, (Business Report for the 36th Year of Operations) (Suwon: Samsung Electronics Corp.).

Shin, J.-S. & Jang, S.-W. (2005) Creating first-mover advantages: the case of Samsung Electronics, SCAPE Working Paper No. 2005/13, Department of Economics, National University of Singapore.

Shindō, T. (2006) Handōtai sangyō no paradaimu shifuto to inobēshon no teitai – senryaku shikō no shien kara mita NEC no konmei no honjitsu (The paradigm shift in the semiconductor industry and the stagnation of innovations – viewing the core of NEC's troubles from a strategy perspective], IIR Working Paper #06-06, Institute of Innovation Research, Hitotsubashi University, Tokyo.

Sung, T.-K. (2005) Giob oi gisul hyeoksinseong gwa gyeolcheong yoin: giob gyumo wha oebu netuwuoku oi yeoghal ul chungsim uro (Determinants of a firm's innovative output: the role of external networks and firm size), *Daehan Gyeongyeong Hakwaechi*, 18(4), pp. 1767–1788.

Tapon, F. & Thong, M. (1999) Research collaborations by multi-national research oriented pharmaceutical firms: 1988–1997, *R&D Management*, 29(3), pp. 219–231.

Tidd, J., Bessant, J. & Pavitt, K. (2005) *Managing Innovation: Integrating Technological, Market and Organizational Change*, 3rd ed. (Chichester: John Wiley).

Triandis, H. C. (1995) *Individualism and Collectivism* (Boulder, CO: Westview Press).

US Department of Commerce (1990) *Japan as a Scientific and Technological Superpower* (Washington, DC: US Government Printing Office).

Wakasugi, R. (1994) Organizational structure and behavior in research and development, in: K. Imai & R. Komiya (Eds) *Business Enterprise in Japan: Views of Leading Japanese Economists*, pp. 159–177 (Cambridge, MA: MIT Press).

Westney, D. E. (1994) The evolution of Japan's industrial research and development, in: M. Aoki & R. Dore (Eds) *The Japanese Firm: Sources of Competitive Strength*, pp. 154–177 (Oxford: Oxford University Press).

Yamaguchi, S. (1994) Collectivism among the Japanese: a perspective from the self, in: U. Kim, H. C. Triandis, Ç. Kâğitcibaşi, S.-C. Choi & G. Yoon (Eds) *Individualism and Collectivism: Theory, Methods, and Applications*, pp. 175–188 (Thousand Oaks, CA: Sage).

Yasumoto, M. & Fujimoto, T. (2005) Does cross-functional integration lead to adaptive capabilities? Lessons from 188 Japanese product development projects, *International Journal of Technology Management*, 30(3/4), pp. 265–298.

Comparing National Innovation Systems in Japan and the United States: Push, Pull, Drag and Jump Factors in the Development of New Technology

Introduction

This essay analyses the relative strengths and weaknesses of the national innovation systems (NIS) in the United States and Japan in supporting sustained new business creation. The study focuses on the national policy histories in the United States and Japan impacting on new firm start-ups, particularly in new technology (R&D type) industries such as life science. These policy histories illustrate the orientation of NIS toward new technology firm start-ups. Tracking global innovation and growth in life science is important; not merely because its advancements can revolutionize human health, but also due to its potential to become the future economic base of leading NIS such as those characterizing the United States and Japan.

Research Questions and Proposition

This study addresses the following research questions:

- What jump-starts technology commercialization, venture capital investment, and new firm formation in new technology industries?
- What are the most effective ways to encourage start-ups and to connect fledgling firms to critical resources?
- What are best practices in national and sub-national (regional) innovation systems in this regard?

In addressing these questions, this study advances the following proposition:

- Push (e.g. policy stimuli), pull (market demand), drag (capital and institutional weaknesses) and jump (targeted community-level strategies) factors underlie the ability of certain locales and countries to create competitive advantages in new technology industries.

Context: The Global Life Science Industry

The global life science industry is currently centred in the United States, which by 2004 supported 75 per cent of the global market. Nearly a third of the world's biotechnology (the largest portion of the life science sector) firms are American (1,444 as of 2005), and the USA dominates in terms of global biotechnology revenues and growth (Ernst and Young Health Sciences, 2005). Japan's life science industry is much smaller in size, a total of 531 firms in 2005.[1] Among university start-ups, a primary source of new business creation in life science, the USA leads with university bio start-ups, with over 400 new university start-ups in 2004 alone. Further, university start-ups have a very high success rate in the United States. To date, over two-thirds of all university start-ups remain in business. In Japan, the total number of university start-ups rose from 1,132 in 2003 to 1,364 in 2004 (232 new) and 1,503 in 2004 (239 new).[2] How national policy in Japan has targeted the life science sector for university-based start up growth is discussed subsequently. Meanwhile, in recent years Japan has been churning out many more patents than the United States, earning it international recognition as the world's most innovative country (*The Economist*, 2007). Japan has yet, however, to see significant growth in life science patents.

The relative potential of Japan to match (and potentially overtake) the United States in certain life science sectors, can be evaluated in terms of key pull, push and jump factors. These include the market (pull); scientific seeds; commercialization (for example via technology licensing organizations or TLOs, university start-ups and incubator support) and other push strategies accelerating the growth of new start-ups out of universities, as well as establishing a venture capital (VC) system. National, regionally-focused policies, or jump factors (for example, supporting new business start-ups) encouraging other local stakeholders to get involved, and the overall sociopolitical and cultural pull of the environment around potential entrepreneurs are also important. This study examines and compares each of these main aspects of the NIS in the United States and Japan in terms of each system's relative strengths and weaknesses. It concludes with a discussion of potential next steps (and missteps) for both countries in maximizing their potential as globally competitive new technology based economies.

Innovation and the Global Life Science Industry: Comparing Developments in Japan and the USA

Market

The market acts as a pull factor in both economies: for example, the baby boomer generation in the USA and the ageing population in Japan have increased demand for health-care products and services, biopharmaceuticals and medical devices. Likewise, the size of the biotechnology market in terms of sales and employment has been growing at a rapid pace in both countries.

Scientific Seeds

At its core, life science is a scientific enterprise, highly dependent on high quality research generating patentable science and technology. This research and development is found primarily in graduate level research programmes and their related institutes at research-oriented universities. Private research laboratories, often funded by large corporate giants (such as pharmaceutical and chemicals firms), as well as top-tier government research laboratories provide additional sources of scientific seeds. The number of patents and scientific papers generated by universities are indicators of this scientific potential. For example, the Graph: University (Life Science) Scientific Publications shows the international rankings of top scientific article generating universities. Harvard University and Tokyo University top the list, followed by the University of California at Los Angeles (UCLA), Michigan and Toronto Universities. The United States dominates 12 of the top 20 spots, with schools including Stanford University, University of California, Berkeley and Johns Hopkins University. Japan is also represented in the top 20, with Kyoto University (7th), Osaka University (15th) and Tohoku University (16th) (see Table 1).

In terms of patents, the following two tables (Tables 2 and 3) list the leading patent generating universities in the USA and Japan. The University of California system (424), California Tech (135) and University of Michigan (132) occupy the top three slots in the United States. The total number of patents is far fewer in Japan, where Osaka University (22) tops the list, followed by Keio University (13) and Tohoku University (11).

Large firms, and especially pharmaceutical and chemical firms, also play a role in generating scientific seeds in the USA. For example, Monsanto, developer of a wide variety of bio-agricultural products (its Roundup brand of herbicide holds a virtual global monopoly), has spawned a number of start-ups, many of which are led by former executives and research staff. While Japanese firms generate as many patents in electronics as American firms, in emerging

Table 1. University (life science) scientific publications: international rankings by number of citations (1991– 99)

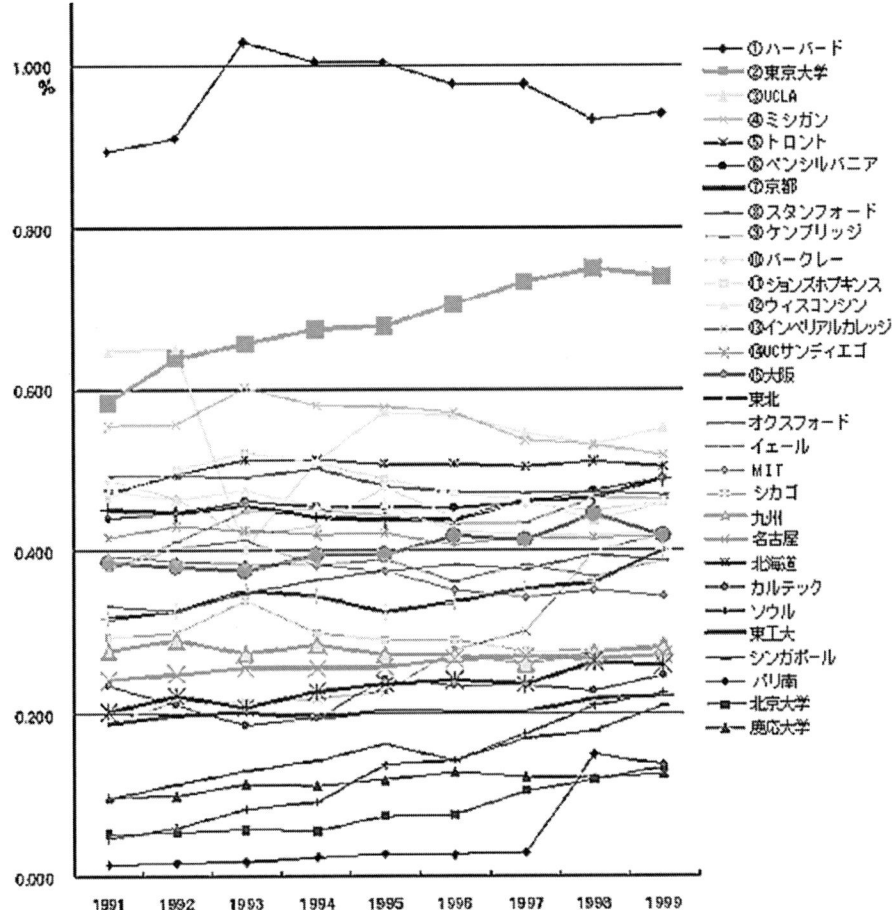

(Ranking: Harvard U., Tokyo U., UCLA, Michigan, Toronto U., Pennsylvania U., Kyoto U., Stanford U., Cambridge U., U.C. Berkeley, Johns Hopkins U., U. of Wisconsin, Imperial College, U.C. San Diego, Osaka U., Tohoku U., Oxford U., Yale U., MIT, U of Chicago, Kyushu U., Nagoya U., Hokkaido U., Cal Tech, Soul U., Tokyo Institute of Technology, Singapore, Paris South, Peking U., Keio U.).
Source: Japan Association of National Universities.

sectors such as biotechnology they lag behind. It has become standard business practice within major Japanese pharmaceutical companies to licence technology developed by American pharmaceuticals rather than investing in their own frontier R&D. Using a play on words, leading Japanese critics compare their system to the role the National Institutes of Health, or 'NIH' plays in the United States in stimulating R&D in emerging sectors such as biotechnology. For these Japanese observers, Japan's 'NIH' is rather, the 'not-invented-here' syndrome.

Table 2. Scientific seeds: top US universities by patents

Rank in 2004	Number of patents in 2004	US university	Start-ups (Rank)
1	424	University of California	5
2	135	California Institute of Technology	14(4)
3	132	Massachusetts Institute of Technology	20(1)
4	101	University of Texas	5
5	94	Johns Hopkins University	5
6	75	Stanford University	9(8)
7	67	University of Michigan	13(5)
8	64	University of Wisconsin	2
9	58	University of Illinois	16(2)
10	52	Columbia University	n/a

Source: Compiled from United States Patent and Trademark Office (2004 data) and the Chronicle of Higher Education's 'Tech Transfer Scorecard' (2005).

Commercialization

TLOs in the USA

The fact that technology licensing organizations (TLOs) have the capacity to generate scientific seeds and subsequently obtain patents does not automatically translate into new firm start-ups, however. For example, universities as a major source of new technology must have the will and wherewithal to commercialize science and technology, either through encouraging faculty to get involved in new firm start-ups or by licensing the technology to existing firms. This function is usually managed by the university technology licensing office and/or related technology licensing organization (TLO). In the United States, the quality of TLOs in terms of commercialization rates (that is, their ability to get university patents licensed, developed and marketed) varies widely. There are approximately 232 university TLOs in the United States.[3]

Contrary to common perception, there is no national US technology licensing model – and thus no model that can be strictly copied – though a few universities do have outstanding TLOs. The epitome of best practice in this regard is the WARF (Wisconsin Alumni Research Foundation) model of the University of Wisconsin, Madison.[4] Established and managed by alumni, WARF was instrumental in licensing the technology to produce Vitamin D in the 1930s, which, in addition to revolutionizing the treatment of rickets (a cause of spinal deformities) in children, netted millions in revenue, much of which has been donated to the University of Wisconsin system. It should be noted that WARF was established completely independently of the university by alumni after university administrators refused to fund the patent application for the vitamin D technology. The WARF model goes one better than a standard TLO, through funding frontier research that might have commercial potential in addition to operating autonomously from the university. More recently, the isolation of human embryonic stem cells in 1998 by Dr James A. Thompson at the University of Wisconsin, Madison was also funded in part by WARF. Though potentially

Table 3. Scientific seeds and commercialization: top Japan universities by patents and start-ups*

２００４年国内大学別特許登録
件数/総合ランキング

順位 (Rank)	出願人 (Name)	件数 (Patents)	ベンチャー企業累積設立数 (# of Firms)	論文被引用数 (Citation)
1	大阪大学 (Osaka U.)	22	71	10.37
2	慶應義塾大学 (Keio U.)	13	50	7.39
3	東北大学 (Tohoku U.)	11	48	7.3
3	東海大学 (Tokai U.)	11		n/a
5	東京大学 (Tokyo U.)	10	92	10.49
6	名古屋大学 (Nagoya U.)	8		8.7
7	金沢工業大学 (Kanazawa Inst. of Tech.)	7		n/a
8	広島大学 (Hiroshima U.)	6		7.14
8	新潟大学 (Niigata U.)	6		8.49
8	東京工業大学 (Tokyo Institute of Tech.)	6	39	6.93
11	九州大学 (Kyushu U.)	5	44	7.85
20	早稲田大学 (Waseda U.)	3	75	n/a
20	京都大学 (Kyoto U.)	3	59	10.21

Note: *Daigaku Hatsu Bencha Ni Kan Suru Kisochosa Jisshi Hokokusho Kabushikigaisha Kachisogokenkyujo.*

lucrative to universities, commercialization of university technology does not automatically translate into new business creation, however. Table 4 shows that, despite the University of Wisconsin, Madison's commercialization success, the university does not even make the top ten nationally in terms of start-ups. Even with those caveats, few other university TLOs have been as successful.

Other top universities in the United States and elsewhere falter at the commercialization stage. This is due in part to university administered TLOs and licensing offices prioritizing maximizing (short-term) university revenue and/or protecting the university from potential liabilities (Toole, 2003; Myers, 2005; Lach & Schankerman, 2004). University licensing offices are often controlled by legal staff and prolonged licensing deal negotiations often sap the life out of potential private sector deals. In sum, the University of Wisconsin, Madison has a TLO that works, while many other top universities have TLOs that do not.

TLOs in Japan

Japan began to acknowledge the role of TLOs in the late 1990s, and by the early 2000s had established a number of TLOs, though success so far has been patchy. The 1998 Law Promoting Technology Transfer from Universities to Industry, (大学等における技術に 関する研 究成果の民間事業者への移転の促進に関する法律)) also referred to as the TLO Law, resulted from a new collaboration of sorts between the Ministry of International Trade and Industry (METI) whose traditional purview includes the small (start-up) business sector, and the Ministry of Education, Science, Sports and Culture (MEXT) which regulates Japanese universities. The TLO Law aimed to encourage universities to get into the technology licensing business, for example by reducing patent fees to government 'approved' TLOs, and dispatching patent experts to local TLOs to assist in the patent application process. Most importantly, the Law gives greater rights to the universities to profit from intellectual property (IP) developed by faculty.

Since new TLOs were eligible to receive government subsidies of JPY20 million for up to five years, it is no surprise that the number of TLOs shot up from less than five in 1998 to 35 in 2003 (Harayama, 2004; Sandelin, 2004, 2007).

Table 4. Commercialization: top ten US universities by start-ups (2004)

Rank	US university	Start-ups
1	Massachusetts Institute of Technology	20
2	University of Illinois	16
3	Georgia Institute of Technology	15
4	California Institute of Technology	14
5	University of Michigan	13
6	Duke University	10
7	University of Pittsburgh	10
8	Stanford University	9
9	University of Colorado	9
10	University of Florida	8

Interestingly, it appears that the most successful TLOs (in terms of commercializing technology and support of new firm start-ups) are not the national public universities but the smaller, and perhaps more nimble (dare we say lower tier?) private universities. In contrast, the coveted 'official government approval' – coveted because of subsequent access to subsidies – of new TLOs still favour the large national universities, which have been historically connected and subsequently well versed in completing government applications and often voluminous amounts of paperwork.

In 1999, following on from the TLO reforms, the Law on Special Measures for Industrial Revitalization (産業活力再生特別措置法) – also called the Industrial Revitalization Law (IRL) – was enacted. The Industrial Revitalization Law was modelled on the 1980 Bayh-Dole Act in the United States, which granted universities patent rights to university faculty developed IP, and likewise, the IRL provided similar rights to Japanese universities. Bayh-Dole and Japan's attempts at implementing a Japanese Bayh-Dole are discussed in the National Policy section below. Before doing so, it makes sense to review the components of a successful entrepreneurial support system that stimulates technology commercialization, including new technology incubators and venture capital – and what they do specifically to help fledgling firms. Getting technology to the marketplace is not limited to commercialization into existing firms. Universities have themselves become hotbeds of entrepreneurial start-ups, particularly in the United States.

University Start-ups

Another way that universities contribute to the growth of new business in emerging sectors, including life science is by encouraging university start-ups. University start-ups are defined here as a new business established using science or technology developed at a university (Di Gregorio & Shane, 2001; Djokovic & Soultaris, 2004).[5] A university faculty may also take an equity stake in the new business, and even be a founder of the new company. In the USA, among the total number of start-ups at leading 'start-up' universities such as MIT and Stanford University, a growing proportion in recent years have been in the biotechnology sector. In the USA, most new technology ventures are established geographically near the universities where the technology originates. Furthermore, in recent years new ventures have become increasingly science-based. For example, among VC-intensive industries in the United States biotechnology has become the leading source of growth in employment, second only to software in sales growth.[6] As the American software industry continues to move offshore, it is expected that the biotechnology industry will become a primary growth engine in the future (See Tables 5 and 6). In Japan, biotechnology start-ups outpace other types, comprising nearly 38 per cent of all university start-ups (a total of 1,112 in 2005), compared to a total of 29.9 per cent in software development.[7]

At universities such as Osaka, Tokyo and Kyoto, more than half of all new start-ups are biotechnology-related.[8] In this regard, METI has set specific benchmarks, aiming for 1,000 university start-ups by the mid-2000s. By 2005 Japan had over

Table 5. Sales growth VC intensive industries 2001–03 (top 5)

	VC sales growth	Total sales growth
Computer software	31%	−2%
Biotechnology	28%	22%
Health-care services	26%	25%
Retailing/media	20%	9%
Computer hardware and services	12%	5%

Source: Venture Impact (2004).

Table 6. Employment growth VC intensive industries 2001–03 (top 5)

	VC employment growth	Total employment growth
Biotech	23%	5%
Computer software	17%	−8%
Retailing/media	12%	−1%
Health-care services	10%	9%
Computer hardware and services	−1%	−14%

Source: Payne (2004).

1,100 university-related start-ups.[9] The proof of success will be in whether or not start-ups survive (bring products to market, profit and grow) in the long term.[10]

Incubators

Another way that university science and technology can be commercialized is through nurturing new university-related businesses within university-sponsored incubators. Like TLOs in both countries, there is wide variation in the quality of incubator facilities. The Small Business Administration (SBA) in the United States was instrumental in encouraging the establishment of incubators in the USA, which in 1980 had only 12 incubators nationwide. Since then, three main types of incubators have been established: university, private sector and government. According to the National Business Incubator Association there are 1,114 business incubators in the USA. Of the 1,400 in all of North America (USA, Canada and Mexico), 25 per cent are university sponsored.[11] It is estimated that between 75 to 90 per cent of incubators in North America are non-profit with an economic development focus.

One distinguishing feature of successful American incubators (measured by 'graduation' rates of tenant firms into independent business status) is that they provide a variety of services above and beyond merely offering firms low cost rental space. These 'extra' services include introductions to the local VC community, access to patent attorneys and accountants (sometimes on a pro-bono basis), and even marketing assistance. Most incubators have a full-time manager whose job it is to support tenant firms and help them grow and eventually graduate out of the incubator and continue on their own. Exemplar incubators also coordinate community-building social events as well, enhancing the potential for a creative, innovative milieu within and around incubators – a milieu comprising member firms and the business and social networks to which the incubator is connected (Wiggins & Gibson, 2003). Studies have found that these quality value-added services have a positive impact on tenant firm performance. According to the National Business Incubator Association (NBIA), of all start-ups the ones that benefit from being in incubators are most likely to stay in business − 87 per cent of incubator tenants are claimed to have survived. Studies have also shown that incubators contribute to the growth of university research parks, often an important conduit between university scientists and the business community (Link & Scott, 2006).

In Japan, the number of incubators is smaller, with the bulk of them being government-sponsored, mostly by city and prefectural governments. By 2002 the Japan Association for New Business Incubation Organizations (JANBO) estimated that there were 325 incubation facilities in Japan. The vast majority of them were established after 1999 after a variety of incentives (for example, subsidies) were put in place by METI and MEXT to encourage incubator formation. According to a survey by JANBO in 2002 of 113 incubators, nearly 80 per cent of incubators were non-profit. In both types of incubators, more than a third of tenant firms were software start-ups.[12] One of the major weaknesses in Japanese incubators is the lack of managerial expertise and other support services provided to tenant firms. In fact, many incubators do not have full-time managers – or, indeed, any managers at all.[13] According to JANBO, less than 10 per cent of incubators surveyed offered tenants support services. 'Support' in this case was usually limited to providing firms with general information (for example, on government programmes). The few incubators that have managers at all tend to be (albeit well-intentioned) career government bureaucrats who have little practical business experience. In a 2005 survey by METI of 371 university related start-up firms, few respondents found incubators to be helpful in any service other than providing the firm space.[14] In contrast, US incubator managers often function as a bridge between firms and venture capital, sales and marketing, and other resources. Even if every Japanese incubator were to be staffed by a qualified IM (incubation manager), Japan still lacks a meaningful number of critical external supports, especially venture capitalists, to which would-be entrepreneurs could be introduced.

Venture Capital

One of the biggest hurdles for a new business start-up is amassing sufficient capital to grow and build business. For science-based start-ups, the initial capital requirements are often much higher than in other sectors such as software owing to the need for laboratory and testing equipment and (often) wet-lab space. Venture capital has played a major role in the United States in supporting new business creation and growth.

The term venture capital or VC describes funds invested in new, unproven businesses. An 'unproven business' is a new enterprise that has no history of sales revenue and profit (in fact, it could merely exist as an idea in the mind of the founder). Broadly, VC is a type of 'private equity' investment in which an equity stake in a firm is taken in exchange for cash investment. In Europe, VC is often referred to as private equity.[15]

Most importantly, venture capital involves hands-on venture management on the part of the venture capitalist. Venture capitalists not only provide money; they also provide relevant know-how and expertise, primarily management, but sometimes also technical. Venture capitalists also provide new entrepreneurs with access to their personal (social) networks, which has benefits for new firms including introducing new members to the board of advisers and qualified service providers. For this reason, individual venture capitalists and venture firms tend to invest mainly in firms in their immediate locales: that is, they tend to be region-specific. Surveys by the National Venture Capital Association (NVCA) confirm that VC firms tend to invest primarily in their immediate locales, and invest in other places as part of a 'syndicate' of investors, where another firm takes the lead investment position – and therefore the greatest risk.

There are generally six stages in VC investment, from the first idea for a new business along to exit, when investors cash out. These are: pre-seed, seed (or 'start-up'), early, expansion, later and exit.

Before a firm becomes a firm, it exists in the minds of the potential entrepreneur: for example, through the discovery, invention of scientific or technological seeds or the development of a new business model. In high-technology industries particularly, new entrepreneurs often seek financial support to flesh out an idea into a product prototype design or to demonstrate that their new product idea has market appeal. This *pre-seed* stage is often described as the one in which the concept for the new business is validated as a good one, called the 'proof-of-concept'.

Once a new entrepreneur has decided (in the best case scenario, after consulting with qualified experts in the area and evaluating the potential market impact and competition in that product space) to go ahead and start a business – he or she starts to put together the people and material resources (infrastructure) of the new firm. This *seed* or *start-up* process usually takes up to 18 months (shorter for software, longer for biotechnology). In the next so-called *early* stage after formation, the firm might be producing prototypes or beta versions of its product, and introducing its product to market. The firm is now usually one to three years into being since inception. By the *expansion* stage, the firm has begun generating sales revenue – though not necessarily profit – and also receiving critical market feedback, helping it to improve its product and expand sales. By so-called *later* stages, the firm has been around (on average) for at least three years and has earned a steady stream of revenue; it may even be profitable. Once a venture firm has entered the later stage of its development, its investors often seek an *exit* – cashing out on their investment. This happens through an initial public offering (IPO), an acquisition of the firm by another firm, re-sale of firm stock to a third party, or a buy-back of equity by the firm's principals. In most cases, the *exit* stage (particularly if the firm is going public) requires a cash infusion: for example, for the services of lawyers and auditors.

In the life sciences, VC has fuelled rapid growth in some of America's stellar start-ups including Genentech. However, a common misperception of the role of VC in the USA attributes the greatest credit to venture capital firms, also referred to as 'classic VC'. This is a misnomer really, since the bulk of venture capital for new firm start-ups is actually of the 'angel' variety. That is, most seed to early stage venture funding comes from high net worth individuals – themselves successful entrepreneurs in many cases. These risk-resilient individuals often become a type of 'saviour' to struggling new firms – thus the label of 'angel' (cf. the study by Tsukagoshi in this current collection). Unfortunately, after the collapse of the dot.com bubble around 2000, classic VC in the USA became risk averse. This is to say, while in biotechnology the amount of funds per investment has risen dramatically, the total number of deals has dropped considerably.

For example, in 2006, the majority of 'classic' or institutional VC went to expansion or later stage firms. At the peak of the VC boom (1999–2000) in the United States, nearly US$95 billion was invested in over 6,000 investments (or 'deals'). After the tech collapse of the 'dotcoms' in the USA, investments fell to US$22 billion by 2002 and the number of investments also experienced a precipitous decline, by over 61 per cent – to just over 2,300 deals. By the year 2006 there were some signs of recovery; however, the United States has yet to return to the peak levels of the year 2000.[16]

This description of the activities of venture capital firms only paints a very small part of the picture of the process of getting a new business from concept-to-market-to-profit. In reality, the so-called classic VC provided by VC firms represents only a tiny proportion of the funds that it takes to get a new firm up and running. To illustrate: according to the Global Entrepreneurship Monitor (GEM), classic VC represents only an average of 13.4 per cent of classic and informal VC put together in 25 GEM countries. In other words, the bulk of start-up capital for new firms comes from informal sources, including the traditional '4 Fs': Founders, Friends, Family and Fools (or foolhardy strangers).[17]

Angels vs. 'Classic' VC

It is estimated that the 250,000 angels in the USA invest in 90 per cent of new firms at their earliest stages of conception. The amounts are not huge – US$2 million or less per investment – but the number of firms influenced by angel investors is significant – upwards of 30,000–50,000 per year.[18] In contrast, of the 1,417 VC firms defined by Dow Jones as 'active' (that is, with at least one investment between 2000 and 2006), most (66 per cent) invested in four or fewer new firms. To illustrate: in 2005 across the entire United States there were a total of only 2,239 VC 'deals' or investments. Furthermore, the angel market is estimated at twice the size of the classic VC market – US$100 billion (angel) versus less than US$50 billion (classic) (Payne, 2004). Table 7 gives a more detailed overview.

Even with these caveats about taking the US VC system as a 'model', some alarming differences can be seen when comparing the state of VC in the USA to other industrial centres, particularly in Japan. Cumulative VC in the United States (that is, venture capital investment that has yet to exit) is the largest in the world, closely followed by cumulative investments in Europe. Far behind are cumulative VC investments in Japan. Tables 8 and 9 give more detail.

VC investment in Japan remains at the lowest rankings of all the 27 Organization for Economic Cooperation and Development (OECD) countries (including the European Union), lags behind Hungary and exceeds only the Slovak Republic.[19] On the positive side, survey research comparing the investment behaviour of Japanese venture capitalists to their American counterparts suggests that Japanese VC investors have lower expectations of

Table 7. Angels v. classic VC in the United States

	# of firms invested per year	Amount per deal	Stage	Market size
Angels 250,000 (SBA)	30–50 K	$2 million or less	90% of seed/start-ups	$100 billion ($30 billion annually)
Venture Capital 1417*	3K	Up to 100s of millions	Expansion/later	$48.3 billion ($ 22 billion annually)

Source: Payne (2004)

return on investment (ROI), and at the same time appear to be more concerned about whether firms are creating new markets (Ray & Turpin, 1990).

National Policy

National policy in both the USA and Japan has attempted to address these capital and other weaknesses. In both countries, national level policies supporting new business creation in emerging sectors have played an important role in facilitating growth in new industries. Apart from signalling national-level support for new businesses in frontier sectors, specific measures have provided incentives and impetus for new business formation. In the United States, for example, the role of SBIRs (Small Business Innovation Research Grants) and STTRs (Small Business Tech Transfer Grants) in the earliest stages

Table 8. Cumulative VC investment in Europe, Japan and the United States (2000–04)

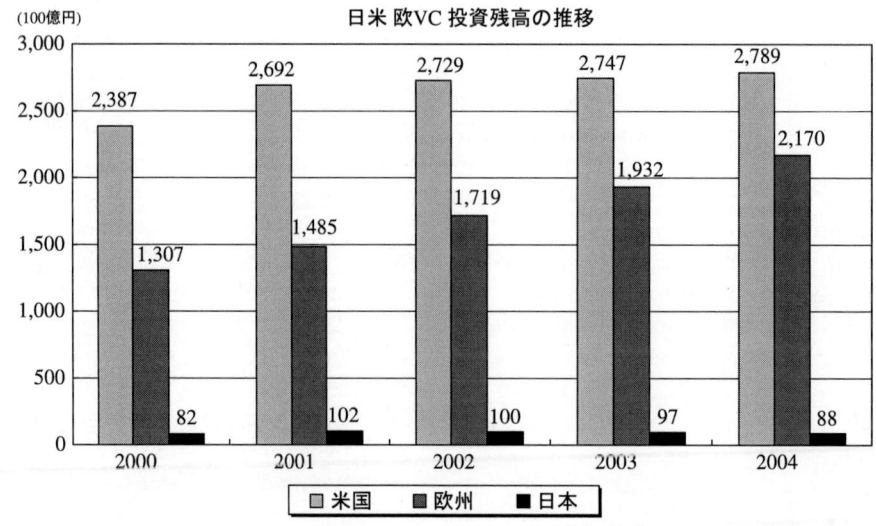

Source: Heisei 17 Nendo Bencha Kyapitaru Nado Toshidoko Chosahokoku, Faundo Benchamaku Hokoku (see note 26)

Table 9. Trends in VC investment in Europe, Japan and the United States (2000–05)

(100億円)   日米欧 VC 投資額の推移

- 2000: 1,132 / 486 / 23
- 2001: 438 / 338 / 28
- 2002: 231 / 384 / 17
- 2003: 202 / 404 / 16
- 2004: 225 / 513 / 15
- 2005: 238 / 653 / 20

□米国   ■欧州   ■日本

(資料) (財) ベンチャーエンタープライズセンター平成17年度ベンチャーキャピタル等投資動向調査報告
および 2005年の米国はNVCA MoneyTree (1$=107円換算)、2005年の欧州はEVCA FinalActivity
Figures 2005 (1ユーロ=139円換算) による。

Source: Heisei 17 Nendo Bencha Kyapitaru Nado Toshidoko Chosahokoku, Faundo Benchamaku
Hokoku (see note 26)

of growth in high technology and science-based new firm start-ups has been
critical in supporting new businesses (Fogarty *et al.*, 2006). Key policies have
included Bayh-Dole (1980), the Small Business Innovation Development Act
(1982 – creating the SBIR Program), the Orphan Drug Act (1983), the Small
Business Tech Transfer Act (1992 – creating the STTR Program), and the
FDA Critical Path Initiative (2004). Reviewing this policy history in the United
States is important because Japan is currently engaged in wholesale copying of
these American national policies in its attempt to create an entrepreneurial
economy.

Bayh-Dole

The intent of Bayh-Dole (1980) in the United States was to establish patent policy
that would encourage patent holders to collaborate with the private sector.
Specifically, the intellectual property rights of inventions resulting from Federal
funding would remain with the inventor under certain conditions. The conditions
included prioritizing small business in the granting of licensing rights to
commercialize technology. The institutions targeted by Bayh-Dole were primarily
federally-funded research institutes, and secondarily universities.

   Since Bayh-Dole was enacted, one somewhat unintended effect has been that
university patents and start-ups have both increased significantly (Nelson, 1998;
Sampat, 2002). Universities have also increased their licensing revenue, though
this has been a long-term process. For example, Stanford University's US$400
million in royalty income between 1991 and 2000 (compared to US$4 million for
the period 1981–90) can be traced to disclosures made back in the 1970s.[20]

SBIRs

The Small Business Innovation Research Program (SBIR) was established in 1982 in order to stimulate public-private sector innovation by requiring 11 major federal departments and agencies to allocate a small percentage of their budgets to American-owned small business (Wallsten, 2000; Audretsch *et al.*, 2002; Cooper, 2003; Gans, 2003; Siegel *et al.*, 2003; Toole & Czarnitzki, 2005; Toole & Turvey, 2005). The largest SBIR granting agencies include the Department of Defense (DoD), the National Science Foundation (NSF) and the Department of Health and Human Services (DHHS, within which the National Institutes of Health, NIH, resides). Awards are granted in two phases: start-up (up to US$100,000) and phase two (up to US$750,000).

Phase one corresponds to the VC proof-of-concept stage, discussed above, whereby funds are granted for about six months in order to test the merit or feasibility of the technology. Phase two awards support further R&D and testing, at this stage aiming for commercialization. Long-term studies comparing SBIR recipient firms to non-recipients have found them to grow faster, and perhaps most significantly, to have had powerful so-called certification effects in attracting private venture capital.[21]

STTRs

The Small Business Technology Transfer (STTR) Program (1992) is similar to the SBIR Program in that the goal has been to promote the commercialization of technology that has been developed with federal funds (Baron, 1993). The main differences are twofold. First, unlike the SBIR, scientists and faculty employed full time at a university and/or research institution are allowed to apply. Secondly, the phase two awards under STTR are currently capped at a lower amount: US$500,000. Also, the number of granting agencies is fewer (only five): Department of Defense (DoD), Department of Energy (DoE), Department of Health and Human Services (DHHS), National Aeronautics and Space Agency (NASA), and the National Science Foundation (NSF). As part of government monitoring of the impact of the aforementioned programmes, the NSF tracks public investment in the national scientific infrastructure, including the SBIRs. Table 10 illustrates the long-term trend in new US patent applications, start-up companies established and trends in licensing revenue.

The Orphan Drug Act

When enacted in 1983, the Orphan Drug Act initially put the onus on firms to demonstrate precisely how cost-prohibitive the R&D expenditure would be in order to develop drugs for the small number of Americans having rare diseases (Rohde, 2000). Start-up firms, however, lack the resources to prepare such time-intensive paperwork; unsurprisingly, until recently few firms applied for orphan drug status. It was not until the Act was revised to allow firms 'Orphan Drug' status if they could demonstrate merely that they were developing a drug for ailments that affected fewer than 200,000 Americans. Major pharmaceutical companies are

Table 10. University licensing activity in the United States (1991–2003)

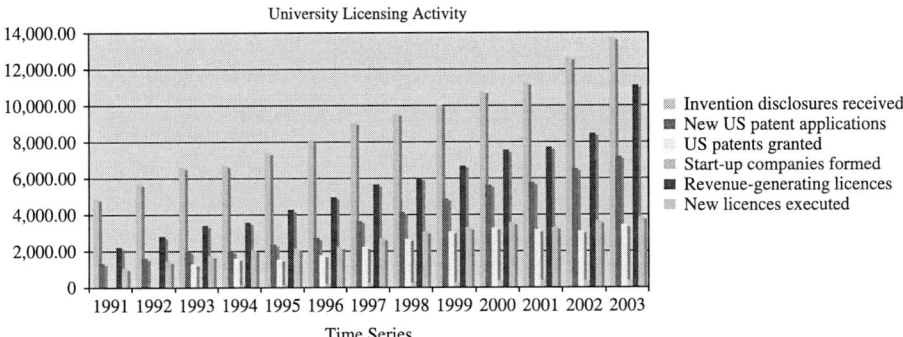

University Licensing Activity

- Invention disclosures received
- New US patent applications
- US patents granted
- Start-up companies formed
- Revenue-generating licences
- New licences executed

Time Series

rarely interested in such a narrow market, making niche market entrance of small firms possible. While orphan drug status does not grant developing firms a patent, it does allow them a seven-year monopoly on the sales of the product. Since its enactment in 1983, a number of drugs have been developed to treat ailments such as tuberculosis. Furthermore, a report by the Department of Health and Human Services noted that orphan drug approval has been helpful in stimulating the development of the biotechnology industry generally: for example, in attracting venture financing (for example, through the certification effect mentioned above) to biotechnology companies developing orphan drugs.[22]

The FDA Critical Path Initiative

In response to the slowdown (during the early 2000s) of new submissions to the US Food and Drugs Administration (FDA) for drugs, therapies and medical device approvals, the FDA published a white paper in 2004 outlining a national strategy to speed up and improve the quality of evaluations of new technologies in the approval pipeline. To date, there is a high failure rate among new prototype products, while at the same time the cost of developing new prescription drugs in particular has risen dramatically – to more than US$800 million by 2000. Furthermore, a higher proportion of potential new products fail in the 2000s than failed in the 1980s, despite major advances in basic science such as in genomics. Consequently, the 2004 FDA Initiative includes measures to improve the evaluation process in terms of the ability to gauge better the likelihood of success in a potential new product. Specifically, measures are under way to utilize better bioinformatics, biomarkers and disease models in evaluating new technologies.[23]

Comparing National Policy Developments in Japan

Japan has emulated several of the aforementioned policies in recent years through its own Bayh-Dole-esque university reform. As discussed above, national universities after 2000 have been expected to fund a significant portion of their own budgets with the intent of having universities act more independently of government, and therefore ideally become more innovative. One result has been to encourage more

private sector initiatives to capitalize on university technology, including supporting
the development of technology licensing and university start-ups. In Japan, the most
significant national policy initiative to date targeting new business creation – apart
from reform of small and medium size enterprise (SME) policy in general – has
been the Innovation Cluster Initiative (Ibata-Arens, 2005).

As I wrote in my book *Innovation and Entrepreneurship* in Japan (2005) and
elsewhere (Ibata-Arens, 2004), METI launched its 'Cluster Initiative' and 'Cluster
Plan' in 2000 and 2001 respectively. The Plan intends to promote innovation and new
business creation, particularly in high technology industries. Related policies by
MEXT (Ministry of Education, Culture, Science, Sports and Technology) are aimed
at encouraging more science and technology-based university start-ups via two main
measures: establishing TLOs (discussed above under the section Commercialization)
and expanding graduate MBA programmes. Within the Cluster Initiative
(whose main growth targets are informatics, biotechnology, nanotechnology and
eco-bio) is an emphasis on promoting the biotechnology industry, particularly in the
Kinki and Hokkaido regions.[24] By fiscal year 2002, the national life science budget
had grown to 440 billion Yen. Other initiatives include the establishment of an SBIR
programme, modelled on the SBIR programme in the USA, as well as measures
targeting improvements in jinzai (personnel skills). The latter includes such activities
as the NEDO Fellow programme, which places young scientists and other
professionals in small businesses – their salaries being paid for a time by the Japanese
government. The most significant change in METI in recent years has been in critical
self-reflection and in-depth survey-based and quantitative benchmark analyses of
how well they are doing in each target area – though some selection bias can be found
in that they tend to survey only firms already involved in METI-sponsored Cluster
Plan projects.[25] Table 11 outlines the main policy changes regarding science-based
new business creation in Japan since the 1990s:

METI has commissioned a number of in-depth academic analyses of their progress
to date (disclaimer: including one in which I am involved). These analyses take
special care to compare commercialization and new firm start-up benchmarks in the
United States and in Japan.[26] One of the major remaining stumbling blocks for Japan,
however, is the form-over-substance problem, alluded to above in the discussion of
the smaller and nimble (private) vs. large and lumbering (public) university attempts
at technology commercialization. That is, while the Japanese government has made
significant steps (particularly after the end of the 1990s) to improve the broader
entrepreneurial climate (or in Japan's case, lack thereof), there remains a significant
drag on national policy stimuli. I have written at length elsewhere about how and why
the post-war Japanese business system that creates such powerful disincentives to
entrepreneurialism: for example, the vertical network integration favouring large,
politically connected organizations (Ibata-Arens, 2005). Other studies have come to
similar conclusions, based on studies specific to university commercialization of life
science research and development (Nilsson *et al.*, 2006).

Comparing Patent Systems

Though reforms began in the late 1990s, essentially the Japanese patent system
was designed to diffuse foreign technologies into large Japanese corporations.

Table 11. Main policy changes for science-based new business creation in Japan

Main policy changes for science-based new business creation in Japan

Year	Policy		URL
1995	Science and technology Basic Law	科学技術基本法	http://www.mext.go.jp/b_menu/shingi/kagaku/kihonkei/kihonhou/point.htm
1995	Basic Plan for Science and Technology	科学技術基本計画	http://www.mext.go.jp/a_menu/kagaku/kihon/honbun/006.htm
1998	Law for Promotion of University-Industry Technology Transfer	大学等における技術に関する研究成果の民間事業者への移転の促進ニ関する法律	http://www.mext.go.jp/a_menu/shinkou/sangaku/sangakuc/sangakuc10_1.htm
1999	Law on Special Measures for Industrial Vitalization (1999)	産業活力再生特別措置法	http://www.meti.go.jp/intro/law/index.html
2000	Law to Strengthen Industrial Technology Capability	産業技術力強化法	http://www.mext.go.jp/a_menu/shinkou/sangaku/sangakuc/sangakuc6_3.htm
2000	Industrial Cluster Plan (Innovation Cluster)	産業クラスター計画	http://www.cluster.gr.jp/relation/data/index.html
2001	Second Science and Technology Basic Plan (2001–05)	第2期科学技術基本計画	http://www.mext.go.jp/b_menu/shingi/gijyutu/gijyutu1/shiryo/001/05061501/11_8.pdf
2003	Basic Law on Intellectual Property	知的財産基本法	http://www.mext.go.jp/a_menu/hyouka/kekka/03082902/040.pdf
2004	National University Law	国立大学法人法	http://www.mext.go.jp/a_menu/koutou/houjin/03052704.htm
2005	Third Science and Technology Basic Plan (2006–10)	第3期科学技術基本計画	http://www.mext.go.jp/b_menu/shingi/gijyutu/gijyutu11/houkoku/05042301.htm

Until the 2000s, the Japanese patent system was geared toward this technology diffusion rather than to the protection of intellectual property. This is evidenced in the policy of 'laying open' to the public the details of all patent applications, immediately after those applications are filed. This has generally resulted in large firms with substantial financial and legal resources to engage routinely in 'patent flooding' of the patents-in-progress approach typical of small firms. In Japan's 'first-to-file' model a shuuhen (surrounding) strategy is used, whereby the larger firm obtains patents on all potential permutations/expansions on the core technology of the original patent. This is in marked contrast to the patent system in the US, which is based on a 'first-to-invent' philosophy. In the latter model, the intellectual property of the inventor holds legal precedence over any later attempts to exploit the invention (Ibata-Arens, 2000). On the other hand, shifts in American sociopolitical attitudes vis-à-vis supporting frontier science have taken away some of the American competitive edge. The study in this current collection (*Asia Pacific Business Review* 14(3)) by Taplin develops this comparison.

Comparing Sociopolitical Cultures

Debates over the development and use of stem cell technology in the USA – fuelled by religious concerns over the use of embryonic stem cells in particular – have put a drag on growth in domestic life science capacity.[27] For example, responding to pressures by Christian lobby groups the Bush administration imposed (in 2001) restrictions on federal funding for stem cell research, limiting federal funds to existing cell lines, that is, those cells already isolated (created) prior to the date of the new restrictions. Further, emboldened by this national signalling, local Christian groups have targeted particular states as 'test cases', aiming for a constitutional amendment forbidding stem cell research. One of these test states is Missouri, home to the emerging St Louis life science cluster.

After a US$30 million public relations campaign, the Stem Cell Initiative (protecting researchers' rights to use stem cells) was narrowly passed via state-wide referendum in November 2006. Of these US$30 million, 29 were funded by Jim and Virginia Stowers, cancer survivors who established the Stowers Institute in Kansas City, Missouri. The Stowers Institute is a leading cancer research centre and stem cell-based therapies are at the forefront of the most promising cancer related R&D. As a result of the 'anti-stem cell' campaign, it has been estimated that several hundred million dollars in research dollars has been lost to other states, notably to researchers at Harvard University – funds that could have been invested in the local economy of Kansas City. Neighbouring states have also tried to capitalize on the controversy in Missouri. In 2005, the governor of Illinois sent a personalized letter to the top 100 scientists in Missouri inviting them to 'come on over'. Rod Blagojevich backed this offer with a state-sponsored initiative of US$10 million to support stem cell research. Meanwhile, California, already the country's leading high technology state, announced a US$10 billion stem cell initiative to be invested over the next ten years.

Competition has also come from farther afield. A 2007 *Business Week* article chronicled the rising incidence of Americans travelling to China to obtain stem-cell based treatments for spinal cord injuries. Clinical developments in China are progressing at a rapid pace and Chinese biotechnology companies are reportedly forging ties with forward thinking others around the world (Einhorn, 2007). What we have in the USA in the early 2000s is a mixed message from the national government, signalling on the one hand support for fast tracking of new drug discoveries, but on the other, indicating that new developments in stem cell therapies should be governed in part by ethical (read 'religious') considerations. Japan's freedom from these politicized drags on the scientific environment for new business creation, might prove a boon in this regard.

Conclusion: Push, Pull, Drag and Jump Factors to Innovation in Japan

Japan has lagged behind the United States in terms of its capacity to produce new technology bases start-ups, particularly in life sciences. However, as outlined in this current study Japan has embarked on a course of catch-up against its closest ally and competitor the United States; notably, in terms of benchmarking improvements in the production of scientific seeds, commercialization and VC development. Japan has also embarked on a revolutionary change of national mindset regarding the role of small-scale businesses and start-ups. For example, one goal is changing the culture for the Japanese national innovation system. Gone are the incremental changes of the muddle-through lost decade of the 1990s. Instead, national government ministries such as METI and MEXT have (in the 2000s) been forced to downplay historical inter-ministry rivalry in order to successfully coordinate their commercialization and entrepreneurship policies. Recent policies, evaluated in this current study, have targeted improvements in the commercialization of science and technology, particularly out of universities – as other studies in this current collection further illustrate. There have been innovation policy shifts in Japan designed to improve the climate for new business start-ups. Table 12 outlines key push, pull, drag and jump factors of the two national innovation systems compared in this study.

Unhampered as it is by a political climate undermining life science development (for example the stem cell debate in the USA) or the financial

Table 12. Life science National Innovation Systems in the USA and Japan: national level push. pull, drag and jump factors

Life science National Innovation Systems in the USA and Japan: national level push, pull, drag and jump factors		
FACTORS	US	JAPAN
Market	Pull	Pull
Scientific seeds	Push	Drag (to push)
Commercialization	Push	Drag (to push)
Venture capital	Push	Drag (to push)
Policy	Push, to drag	Push (to jump)
Patent system	Push	Drag**
Sociopolitical culture	Entrepreneurialism and pull	Anti-entrepreneurial drag to Pro-entrepreneurial culture

drain of a war in the Middle East, Japan's remaining anti-entrepreneurial drag might just transform into something else. Ideally, this transformation will result in a jump towards the creation of a more pro-entrepreneurial culture, supported by the national government and manifested in the nurturing of globally competitive new technology businesses in Japan – implications of which are discussed in other studies presented in this collection.

Notes

[1] Japan Bioindustry Association (bioindustri kyokai) (2006). See p. 6 graph for growth trend 1994–2000. In Japan, what is referred to as 'bio' is what we call 'life science'. That is, in Japan, 'bio' encompasses biopharmaceuticals, medical devices, bio-informatics, eco-bio and the like. See Biovencha sokeichosahokokusho (2005) p. 5, Table 9.

[2] Report on Basic Survey on University Ventures (2005) (Tokyo: METI), p. 5.

[3] Autm Survey 2004 (Deerfield, IL: The Association of University Technology Managers).

[4] Science and Technology in Congress (Washington, DC: Center for Science, Technology, and Congress, 2001).

[5] More than 3,000 university start-ups were in existence by 2004, according to the AUTM Survey.

[6] A lot of venture capital investment in a certain sector is an indicator of high growth potential of that sector.

[7] Report on Basic Survey on University Ventures, p. 9.

[8] Ibid. pp. 10, 18.

[9] Survey Report on Support for Success in University Ventures, January 2006 (Tokyo: METI).

[10] Kirihata has studied the challenges facing new technology firm start-ups in Japan in the context of recent MEXT policies supporting new business creation. See both Kirihata's undated publications in the References.

[11] See Business Incubation Faq, National Business Incubator Association. Available at http://www.nbia.org/resource_center/bus_inc_facts/index.php (accessed 14 July 2007).

[12] Evaluation Report on Status of Incubation Infrastructure (Inkyubeshan shisetsu no jittai chosa), February 2002 (Tokyo: METI).

[13] According to METI, Japan has a much lower ratio of incubation managers to incubators than the U.S. The Current Direction and Strategy of Japanese Business Incubators, June 2005, (Industrial Infrastructure Department, Regional Economy Industrial Group (Tokyo: METI).

[14] Report on Basic Survey on University Ventures, Tokyo.

[15] In the United States, private equity sometimes has a negative image, as private equity firms tend to seek to 'flip' their investments and turn a quick profit, for example, by lancing out faltering parts of the company. This tactic, is particularly reviled by labour groups, as it often leads to significant lay-offs.

[16] 2006 Venture Capital Industry Report, (Dow Jones Venture Source, 2006).

[17] 2005 Executive Global Entrepreneurship Monitor (GEM) Report, Babson College.

[18] Angel finance is often informal, that is, non-contractual, in nature. Consequently, published figures underestimate total angel finance.

[19] Venture Capital Investment by Stages as a Share of GDP, 1999–2002, Science Technology Industry, Venture Capital: Trends and Policy Recommendations (Office of Economic Co-operation and Development, 2003).

[20] See Sandelin (2004). Some have cast doubt regarding the true impact of Bayh-Dole, for example by arguing that while the number of patents increased after Bayh-Dole, the quality of patents declined. Sampat *et al.* (2003) re-examine this thesis using longer-term patent data. See also Mowery *et al.* (2001); Mowery and Sampat (n.d.).

[21] See Lerner (1996). Annual Reports are available from the U.S. Small Business Administration that track the kinds of investments that the SBIR Program makes.

[22] The Orphan Drug Act: Implementation and Impact, ed. Office of the Inspector General Department of Health and Human Services (Department of Health and Human Services, May 2001).

[23] Challenge and Opportunity on the Critical Path to New Medical Products: Innovation or Stagnation?, ed. Department of Health and Human Services (U.S. Food and Drug Administration, 2004).

24 Sangyo Kurasuta Kenkyukai Hokokusho Heisei 17 Nen 5 Gatsu Sangyo Kurasuta Kenkyukai, (Industrial Cluster Group, May 2005).

25 Sangyo Kurasuta Keikaku Monitaringu Chosa Hokokusho Heisei 17 Nen 3 Gatsu Kabushikifaisha Mistubishi Sogokenkyujo, (Mitsubishi Research Institute, March 2005); Heisei 17 Nendo Sangyo Kurasuta Keikaku Monitoringu Nado Chosa Hokokuso Heisei 18nen 3 Gatsu, Kabushikigaisha, Ribetasu Connsarutingu, (Libertas Consulting, March 2006); Sangyo Kurasuta Kenkyukai Hokokusho Heisei 17 Nen 5 Gatsu Sangyo Kurasuta Kenkyukai (Industrial Cluster Group, May 2005); Sangyo Kurasuat Kenkyukai Hokokusho (Gaiyo), (Industrial Cluster Study Group, 2005); Sangyo Kurasuta Dai Ni Ki Chukankeikau, Heisei 18 Nen 4 Gatsu 1 Nichi, Sangyokeisaisho, Chiiki Keizai Sangyo Gurupu (Regional Economic and Industrial Policy Group, METI, 2006).

26 Daigaku Hatsu Bencha Ni Kan Suru Kiso Chosa Hokoku Sho Heisei 17 Nen 6 Gatsu (METI); Daigaku Hatsu Bencha Ni Kan Suru Kisochosa Jisshi Hokokusho Kabushikigaisha Kachisogokenkyujo (Value Management Institute); Daigaku Hatsu Bench No Seishoshien Ni Kan Suru Chosahokokusho Heisei 18 Nen Ichi Gatsu Keizaisangyosho (METI, 2006); Inkyubeshon Shisetsu No Jittai Chosa 2002 Nigatsu (Japan Association of New Business Organizations, 2002); Wagakunino Kongo No Bi Shisaku Hokosei to Senryaku Nitsuite Heisei 17 Nen 6 Gatsu Keizaisangyosho Chiikikeizai Sangyo Gurupu Sangyo Shisetsuka, (Regional Economy Industrial Group, METI, June 2005); Heisei 19 Nen Yosan Gaisan Yokyu Nado Ni Kakawaru Jimae Hyokasho Heisei 18 8 Gatsu, (METI, 2006); Kenkyu Kaihatsu Gata Bencha to Shien Senmonka Tono Senraku Teikei, Kenkyujigyo Hokokusho, (METI); Heisei 17 Nendo Bencha Kyapitaru Nado Toshidoko Chosahokoku, Faundo Benchamaku Hokoku, (Venture Enterprise Center, 2005).

27 Stem cells are of two types: adult and embryonic. The debate over stem cell research is over the use of embryonic, though in public discourse the two have often been conflated.

References

Association of University Technology Managers (2004) Autm Survey (Deerfield, IL: Association of University Technology Managers).

Audretsch, D. B., Link, A. N. & Scott, J. T. (2002) Public/private technology partnerships: evaluating SBIR-support research, *Research Policy*, 31(1), pp. 145–158.

Baron, J. (1993) The Small Business Technology Transfer (STTR) Program: converting research into economic strength, *Economic Development Review*, 11(4), pp. 63–66.

Biovencha sokeichosahokokusho (2005) Tokyo.

Center for Science, Technology, and Congress (2001) Science and Technology in Congress (Washington, DC: Center for Science, Technology, and Congress).

Cooper, R. S. (2003) Purpose and performance of the Small Business Innovation Research (SBIR) Program, *Small Business Economics*, 20, p. 14.

Di Gregorio, D. & Shane, S. (2001) Why do some universities generate more start-ups than others?, *Research Policy*, 32(2), pp. 209–227.

Djokovic, D. & Soultaris, V. (2004) Spinouts from academic institutions: a literature review with suggestions for further research (London: Cass Business School).

Dow Jones Venture Source (2006) 2006 Venture Capital Industry Report.

Economist, The (2007) Economic and financial indicators, 19 May, pp. 105–106.

Einhorn, B. (2007) Stem-cell refugees: Yanks are flocking to China for therapy, *Business Week*, 12 February.

Ernst and Young Health Sciences (2005) Beyond Borders: Global Technology Report 2005.

Executive Global Entrepreneurship Monitor (GEM) (2005) Report, Bobson College.

Fogarty, M. S., Sinha, A. K. & Jaffe, A. B. (2006) ATP and the US innovation system: a methodology for identifying enabling R&D spillover networks, National Institute of Standards and Technology. Available at http://www.atp.nist.gov/eao/gcr06-895/gcr06-895_report.pdf (accessed 20 March 2008).

Gans, J. S. & Stern, S. (2003) When does funding research by smaller firms bear fruit?: evidence from the SBIR Program, *Economics of Innovation and New Technology*, 12(4), pp. 361–384.

Harayama, Y. (2004) Japanese technology policy on technology transfer: development of technology licensing organizations and incubators, *Tech Monitor*, 6 (March-April), pp. 30–36.

Ibata-Arens, K. (2000) The business of survival: small and medium sized high tech firms in Japan, *Asian Perspective*, 24(4) (Special issue on Dysfunctional Japan, ed. C. Johnson), pp. 217–242.

Ibata-Arens, K. (2004), Japan's quest for entrepreneurialism: The Cluster Plan, Japan Policy Research Institute Working Paper No. 102 (August).

Ibata-Arens, K. (2005) *Innovation and Entrepreneurship in Japan: Politics, Organizations and High Technology Firms* (Cambridge: Cambridge University Press).

Industrial Cluster Group (2005) Sangyo kurasutakenkyukai hokokusho (gaiyo) (Tokyo: Industrial Cluster Group).

Industrial Cluster Study Group (2005, May) Sangyo Kurasuta Kenkyukai Hokokusho Heisei 17 Nen 5 Gatsu Sangyo Kurasuta Kenkyukai (Tokyo: Industrial Cluster Group).

Japan Association of New Business Organizations (2002) Inkyubeshon shisetsu no jittai chosa 2002 nigatsu (Tokyo: JANBO).

Japan Association of National Universities (n.d.) 図表1-1世界の自然科学発表論文数にしめる各国主要大学のシェア社団法人 国立大学協会. Available at http://www.kokudaikyo.gr.jp/active/txt6-1/h14_5/01.html.

Japan Bioindustry Association (bioindustri kyokai) (2006) 2005 Nen Bio Vencha Sokeichosahokokusho.

Kirihata, T. (n.d.) MOT to Overcome the Valley of Death for New Technology-Based Small Firms in Japan, No. 25 (Kansai Science City: Nara Institute of Science and Technology).

Kirihata, T. (n.d.) What Should the Cluster Policy Do to Foster New Technology Based Small Firms? The Case of the Intellectual Cluster Policy of Mext in Japan, No. 11 (Kansai Science City: Nara Institute of Science and Technology).

Lach, S. & Schankerman, M. (2004) The impact of royalty sharing incentives on technology licensing in universities, *Journal of the European Economic Association*, 2(2-3), pp. 252–264.

Lerner, J. (1996) The government as venture capitalist: the long-run impact of the SBIR Program, No. 43, National Bureau of Economic Research.

Libertas Consulting (2006, March) Heisei 17 nendo sangyo kurasuta keikaku monitoringu nado chosa hokokuso Heisei 18 nen 3 gatsu, Kabushikigaisha, Ribetasu Connsarutingu.

Link, A. N. & Scott, J. T. (2006) U.S. university research parks, *Journal of Productivity Analysis*, 25(12), pp. 43–55.

METI (Ministry of Economy, Technology and Industry) (2002) Evaluation Report on States of Incubation Infrastructure (Inkyubeshon shisetsu no jittai chosa) (Tokyo: METI).

METI (2005) Basic Survey on University Ventures (Tokyo: METI).

METI (2005) Kenkyu kaihatsu gata bencha to shien senmonka tono senraku teikei, Kenkyujigyo hokokusho, Heisei 17 nendo bencha kyapitaru nado toshidoko chosahokoku, faundo benchamaku hokoku (Tokyo: METI).

METI (2006) Heisei 19 nen yosan gaisan yokyu nado ni kakawaru jimae hyokasho Heisei 188 gatsu (Tokyo: METI).

METI (2006) Daigaku hatsu bench no seishoshien ni kan suru chosahokokusho (Tokyo: METI).

METI (2006) Survey Report on Support for Success in University Ventures (Tokyo: METI).

Mitsubishi Research Institute (2005, March) Sangyo kurasuta keikaku monitaringu chosa hokokusho Heisei 17 nen 3 gatsu Kabushikifaisha Mitsubishi sogokenkyujo (Tokyo: Mitsubishi Research Institute).

Mowery, D. C., Nelson, R. R., Sampat, B. N. & Ziedonis, A. A. (2001) The growth of patenting and licensing by U.S. universities: an assessment of the effects of the Bayh-Dole Act of 1980, *Research Policy*, 30, p. 20.

Mowery, D. C. & Sampat, B. N. (n.d.) Universities in National Innovation Systems (Berkeley, CA: Class School of Business, UCLA & School of Public Policy, Georgin Institute of Technology.

Myers, R. A. (2005) Challenges for Japanese universities' technology licensing offices: what technology transfer in the United States can tell us, in: Center on Japanese Economy and Business Working Paper Series (New York: Columbia University), pp. 1–38.

Nelson, L. (1998) The rise of intellectual property protection in the American university, *Science*, 6 (March), pp. 1460–1461.

Nilsson, A. S., Friden, H. & Schwaag Serger, S. (2006) Commercialization of life-science research at universities in the United States, Japan, and China, Research Paper No. 100, Stockholm: Swedish Institute for Growth Policy Studies.

Office of Economic Co-operation and Development, (OECD) (2003) Venture Capital Investment by stages as a share of GDP, 1999–2002, Science Technology Industry, Venture Capital: Trends and Policy Accommodations (Washington, DC: OECD).

Office of the Inspector General Department of Health and Human Services (2001) The Orphan Drug Act: Implementation and Impact (Washington, DC: Department of Health and Human Sciences).

Department of Health and Human Services (2004) Challenge and Opportunity on the Critical Path to New Medical Products: Innovation or Stagnation? (Washington, DC: US Food and Drug Administration).

Payne, W. F. (2004) Angels Shine Brightly for Start-up Entrepreneurs, in: *Kauffman Thoughtbook* (Kansas City: Kauffman Foundation).

Ray, D. M. & Turpin, D. V. (1990) Factors influencing Japanese entrepreneurs in high-technology ventures, *Journal of Business Venturing*, 5(11), pp. 91–102.

Regional Economy Industrial Group (2005, June) Wagakunino kongo no BI shisaku hokosei to senryaku nitsuite Heisei 17 nen 6 gatsu Keizaisangyosho Chiikikeizai sangyo gurupu sangyo shisetsuka (Tokyo: METI).

Regional Economic and Industrial Policy Group (2006) Sangyo kurasuta dai ni ki chukankeikau (Tokyo: METI).

Rohde, D. D. (2000) The Orphan Drug Act: an engine of innovation? At What Cost?, *Food and Drug Law Journal*, 55(23), pp. 125–148.

Sampat, B. N. (2002) Private parts: patents and academic knowledge in the twentieth century. Paper prepared for AAAS/CSPO Research Symposium for the 'Next Generation' of Leaders in Science and Technology Policy, 22–23 November, Washington, DC.

Sampat, B. N. & Mowery, D. C. (n.d.) Universities in National Innovation Systems. Available at http://www. globelicsacademy.net/pdf/DavidMowery_1.pdf (accessed 20 March 2008).

Sampat, B. N., Mowery, D. C. & Ziedonis, A. A. (2003) Changes in university patent quality after the Bayh-Dole Act: a re-examination, *International Journal of Industrial Organization*, 21(19), pp. 1371–1390.

Sandelin, J. (2004) Japan's industry-academic-government collaboration and technology transfer practices: a comparison with United States practices, *Journal of Industry-Academia-Government Collaboration*, 3(4), pp. 1–4.

Sandelin, J. (2007) Universities and industries exchange technologies in America and Asia, in: M. Gong Hancock, H. S. Rowen & W. F. Miller (Eds) *Making It: The Rise of Asia in High Tech*, pp. 323–349 (Stanford, CA: Stanford University Press).

Siegel, D. S., Wessner, C., Binks, M. & Lockett, A. (2003) Policies promoting innovation in small firms: evidence from the US and UK, *Small Business Economics*, 20(6), pp. 121–127.

Toole, A. A. (2003) Understanding entrepreneurship in the US Biotechnology Industry: characteristics, facilitating factors, and policy changes, in: D. M. Hart (Ed.) *The Emergence of Entrepreneurship Policy: Governance, Start-Ups, and Growth in the U.S. Knowledge Economy*, pp. 175–194 (New York: Cambridge University Press).

Toole, A. A. & Czarnitzki, D. (2005) Biomedical academic entrepreneurship through the SBIR Program. NBER Working Paper No. W11450. Available at http://papers.ssrn.com/sol3/papers.cfm?abstract_id=755687 (accessed 20 March 2008).

Toole, A. & Turvey, C. (2005), The relationship between public and private investment in early-stage biotechnology firms: is there a certification effect? Paper presented at an International Conference on Agricultural Biotechnology, 6–10 July, Ravello, Italy

Value Management Institute (2005) Daigaku hatsu bencha ni kan suru kisochosa jisshi hokokusho kabushikigaisha Kachisogokenkyujo (Tokyo: Value Management Institute).

Venture Impact (2004) Venture Capital Benefits to the US Economy (Arlington, VA: National Venture Capital Association).

Wallsten, S. J. (2000) The effects of government-industry R&D Programs on private R&D: the case of the Small Business Innovation Research Program, *The RAND Journal of Economics*, 31(1), p. 18.

Wiggins, J. & Gibson, D. V. (2003) Overview of U.S. incubators and the case of the Austin Technology Incubator, *International Journal of Entrepreneurship and Innovation Management*, 3(10), pp. 56–65.

Growing R&D Collaboration of Japanese Firms and Policy Implications for Reforming the National Innovation System[1]

KAZUYUKI MOTOHASHI

Introduction

One characteristic of Japan's national innovation system is the pivotal role played by large corporations and, compared with the USA, the relatively small role played by small firms. According to the R&D Survey conducted by the Japanese Statistical Bureau, the total R&D expenditure of corporations was 11.8 trillion yen in 2003. According to a Nikkei Newspaper Survey, the R&D budget of the top ten corporations for 2004 totalled 4.4 trillion yen, which is almost 40 per cent of the total R&D expenditure in the private sector. R&D activity is concentrated in a small number of large firms.

Owing to the relatively abundant in-house R&D resources at large firms, they are reluctant to R&D collaborations including ones with universities in general. This type of 'in-house preference' exhibited by large enterprises is sometimes referred to as an NIH (Not Invented Here) syndrome, which can also be observed in the USA. However, the in-house orientation of R&D activities could be better explained by the Japanese-style economic system, typified by bank-centred capital markets and corporate government systems, and long term employment practices. Due to the lack of active researcher turnover, R&D collaborations across firms and universities are hampered. Moreover, due to the bank-centred capital market, the mechanism for supplying risk money that should naturally be undertaken by venture capitalists is feeble. Therefore, large corporations play a dominant role in R&D using their internal financial resources, while the activities of start-up firms are not so strong (Motohashi, 2001).

Recently, however, there are signs of change; large corporations are now eagerly seeking for opportunities in R&D collaborations. According to the R&D Collaboration Survey conducted by Research Institute for Economy, Trade and Industry (RIETI) in February 2004, firms are treating external collaborations more positively compared with five years ago, regardless of industries and firm size (RIETI, 2004). In an era of globalization and innovation competition, spurred by the catching up Japan's neighbor countries such as Korea and Taiwan, it becomes increasingly difficult for Japanese large corporations to conduct all R&D in-house. The increasing importance of scientific knowledge in the R&D process of enterprises in certain industries, such as pharmaceuticals, is also a factor fostering external collaborations, particularly with universities and public research institutions (Motohashi, 2005b).

The Emerging Significance of NTBFs and UICs

As Japan's innovation system shifts from large firm in-house R&D systems towards network-based systems based on external collaborations, new technology-based firms (NTBFs) – new, relatively small firms seeking to commercialize innovative ideas – will play a key role. According to several recent surveys, small firms are more active in external collaborations due to the lack of in-house management resources (RIETI, 2003, 2004). Moreover, as large Japanese corporations grope for a new style of R&D management based on external collaborations, NTBFs become important partners for these collaborations. Large enterprises that conduct business globally have been actively collaborating with foreign start-up firms, but language and geographic barriers have impeded effective collaboration in such activities. Therefore, fostering competitive NTBFs will become a crucial issue for Japan's innovation system in an era of R&D networks.

Another key factor in the reform of Japanese innovation system is the vitalization of university industry collaborations (UICs). A series of legislative actions are relevant here: for example, the Technology Licensing Organization (TLO) Promotion Law enacted in 1998; the Law to Strengthen Industrial Technological Capabilities implemented in 2000; and a change in the organizational form of national universities in 2004, which has enabled these universities to become independent of government. These policies have helped swiftly to consolidate a basis to promote UIC activities, while the institutional changes have spurred the interests of private companies to get involved in UIC activities. However, although UIC activities used to play a limited role in Japanese innovation systems, small firms have been more successful in implementing UIC activities by being effective in crossing the organizational boundaries between corporations and universities (Kodama, 2003).

This study examines R&D collaboration activities of SMEs, focusing on UIC activities and their role in changing the Japanese innovation system. First, the state of external collaborations of R&D in Japan is discussed, including small firms. For this discussion data is used from the R&D Collaboration Survey by RIETI. This is a detailed survey that covers the partners involved in the external collaboration, such as large corporations, small or start-up firms, universities, and public research institutes. In this way the differences between UIC activities performed by small firms and other forms of external collaborations can be juxtaposed while the role of small or start-up firms as a partner for external collaborations is examined.

The results are then provided of the quantitative analysis that examines the R&D productivity of firms with a focus on UIC activities conducted by small firms. Motohashi (2004) and Motohashi (2005a) report that small firms have utilized UIC activities more effectively than large enterprises using data on the UIC Survey conducted by RIETI in 2003, but this issue is further examined using detailed data such as the contents of the collaboration. In the final chapter, the results of this study are summarized and discussed in the context of the Japanese national innovation system changing from an in-house system dominated by large corporations to a network-based one.

Research Proposition

Overall, this study advances the following proposition:

- In Japan, self-contained large companies drive the national innovation system and the role of start-up companies remains relatively minor. However, this is not an appropriate system in a world of increasingly open innovation.

In advancing this proposition, this essay explores evidence that the current situation for innovation in Japan is changing, as exemplified by changing patterns of R&D collaboration between Japanese firms, notably in relation to small and medium sized enterprises (SMEs) and start-up firms. This study highlights some of the implications generated by these changing patterns of collaboration between firms for processes of reforming Japan's national innovation system (NIS).

Context: The Role of SMEs in R&D Collaborations

There are several studies and surveys covering UIC activities in Japan. For example, the Mitsubishi Research Institute (2002) conducted a survey at universities, and Ministry of Economy, Trade and Industry (METI) (2003a)), METI (2003b), and Japan Finance Corporate of Small and Medium Size Enterprises (JASME) (2002) have conducted surveys focusing on firms. In addition, METI (2001) conducted a survey of both enterprises and universities participating in UIC activities receiving subsidies from the New Energy and Industrial Technology Development Organization (NEDO) and collected data from both sides of the collaboration. These surveys reveal the attitudes, awareness of problems, the actual problems and the environment of UIC activities from both perspectives. In addition, one can refer to a study that has examined UIC activities by investigating the number of companies involved, the geographical expansion of such efforts, and the number of projects by technology, focusing on UIC activities that have received public grants (Wen & Kobayashi, 2001). However, these examples provide only qualitative information concerning UIC activities: quantitative information such as annual budgets for collaborative research or the number of consigned research contracts are not collected. Furthermore, these surveys are limited in scope, that is, targeting only UIC activities with public grants or focusing on only large firms.[2]

RIETI's R&D Collaboration Survey

In contrast, RIETI's R&D collaboration is based on a list of patent applicants at the Japanese Patent Office and covers almost all firms conducting R&D activities. A survey was conducted of 5,000 firms drawn from a list of firms that submitted more than three patent applications in 2001; 556 valid responses were collected. Since the survey on 175 out of the 5,000 firms could not be conducted due to factors such as unknown addresses, the sample size is 4,825 with an effective response rate of 11.5 per cent. The survey covers a broad range of items, including trends and background factors that lead to external collaborations in R&D, the use of licences for patents, changes in the environment surrounding R&D activities, and the scope and extent of external collaborations.[3]

Trends in External R&D Collaborations

From among the survey respondents, it is found that more than 70 per cent of firms were engaged in some form of external collaboration with other firms, universities, or public research institutes. In this survey, firms are asked about current activities as well as the situation five years previously. The results show that collaborations with large corporations have increased from 31.2 per cent to 37.5 per cent, collaborations with small firms have increased from 22.2 per cent to 38.7 per cent, and collaborations with universities have increased from 39.7 per cent to 51.3 per cent, thus recording an increase in the ratio of external collaborations for all partners. These results reveal that the R&D process in Japan has been shifting from an in-house oriented system to a network based one. Furthermore, out of the different types of partners, more than half of the firms surveyed participate with universities; the ratio of firms collaborating with universities has increased most rapidly.

Let us now take a closer look at external collaborations with respect to firm size. Figure 1 shows the proportion of the firm's R&D collaborations with respect to the firm's total R&D (the share in the R&D budget) grouped by firm size. First, the larger the size of the firm, the higher is the proportion of firms engaging in some form of R&D collaboration. Approximately half of the small firms have not

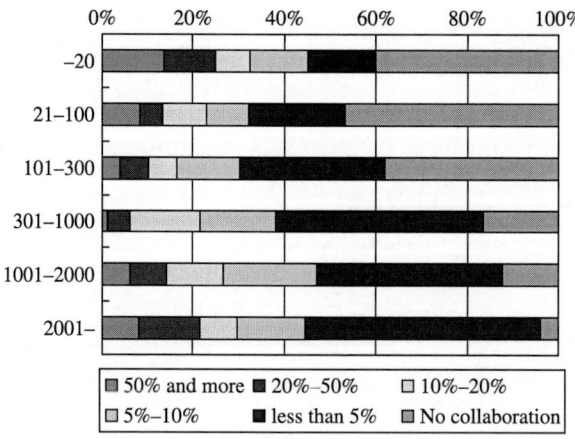

Figure 1. Share of outsourced R&D to total R&D by firm size

engaged in any form of external collaboration, while most of the firms with more than 2,000 employees have done so. When one looks at trends in the proportion of R&D collaborations with respect to total R&D, the proportion is higher for smaller firms. For example, out of the firms with fewer than 20 employees, more than 20 per cent have R&D collaboration ratios higher than 50 per cent. The proportion of these firms declines until one reaches those in the category with between 301 and 1,000 employees; it then increases after this threshold, exhibiting a U-shaped pattern.[4]

The smallest category of firms comprises mainly high-technology start-up firms that do not possess sufficient resources to conduct in-house R&D, and this may explain why these firms tend to engage actively in R&D collaborations. Nevertheless, additional in-house R&D is essential in order to make use of R&D collaborations rather than simply introducing the technology from external parties. If the technological capacity or the absorptive capacity of the firm is critical, as in the aforementioned case, then it will be more effective for larger enterprises to conduct external collaborations (Cohen & Levinthal, 1990). The results of Figure 1 could be interpreted as demonstrating both these effects. Several studies have investigated the effects of UIC activities with respect to firm size, mainly in the USA. For example, Cohen *et al.* (2002b) have shown that large enterprises were more active in UICs than SMEs.[5] In contrast, Acs *et al.* (1994) showed that for innovation activities such as new product introductions, SMEs utilized the results of university research more effectively, while companies with poor in-house R&D resources tended to be more active in utilizing external resources. In Figure 1, both of these factors are shown up behind the U-shaped relationship between firm size and the intensity of R&D collaborations.

Examining these differences in more detail, Figure 2 shows the survey results of the effects of R&D collaborations with SMEs and large firms, grouped by firm size. The ratios in the figure correspond to the proportion of firms that reported positive effects for each type of effect. First, the effects with the highest proportions for firms with fewer than 20 employees are 'new R&D project', 'R&D speed', and 'new product development'. In medium-sized firms (that is, with

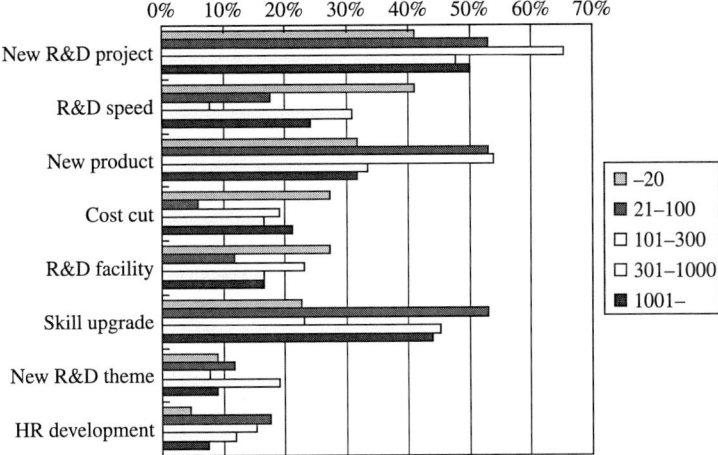

Figure 2. Effects of R&D collaboration by firm size

between 21 and 300 employees) the proportions of 'new R&D project' and 'new product development' are high. For firms with more than 300 employees, 'new R&D projects' and 'skill upgrading' are relatively high.

Small firms have a limited amount of resources for R&D, and their desire to compensate for these deficiencies via external collaborations may be reflected in these results. For example, high technology start-ups, which have not yet gained substantial sales, need to speed up the R&D process. Next, the proportion of external collaborations with respect to total R&D is relatively small for medium-sized firms, as can be seen from Figure 1, reflecting their strategy of engaging in R&D that leads to new products with a relatively small scale of external collaboration. Finally, for large firms, the share of R&D collaboration that leads to final products decreases, for these may be focusing on R&D collaborations with long-term effects that will increase the firm's in-house R&D capacity. Therefore, the relevance of 'skill upgrading' is relatively higher for this group. This type of analysis reminds us that external collaborations differ greatly depending on the specific conditions of the company, and to realize the existence of heterogeneity regarding R&D collaborations.

UIC activities by Size of Firm: Results from Recent Research

A detailed explanation of the differences in firms' attitudes toward university industry collaborations (or UICs) contingent on firm size is given in Motohashi (2004) and Motohashi (2005a). The following are some key conclusions from these two studies:

- The larger the size of the firm, the larger the proportion of the firm's engagement in UIC activities. However, the gap has shortened over the past five years.
- Large enterprises tend to target joint research that will lead to benefits in the long run such as upgrading one's R&D potential, while small firms largely focus on UIC activities that lead to new product development.

- As for the problems associated with UICs, a large proportion of small firms listed 'inadequate experience by the firm', while large enterprises mainly listed reasons related to contractual issues such as 'unclear responsibilities' and 'unclear contract'.

As the points above indicate, the attitudes toward UICs differ between small and large enterprises, and the types of problems faced differ depending on the size of the firm. Research conducted at universities covers a wide variety of disparate areas, and various forms of UICs exist ranging from technology consulting to exchange of researchers as well as joint R&D. The form of collaboration is contingent upon the specific situation of the firm. In the survey conducted in 2004, similar results have been reported with regard to the proportion of firms engaged in R&D collaborations and to their objectives for engaging in UICs, grouped by firm size. As well as the aforementioned points, the current survey asks firms about the approximate expected duration before the results of the R&D collaboration become commercialized products. Figure 3 shows the summary of the responses for this question grouped by firm size.

The proportion of firms focusing on relatively short-term effects is high for small firms, while the proportion of responses indicating longer-term merits increase with respect to firm size. For firms with fewer than 20 employees, close to 20 per cent of these expect commercialization of products within one year, and over 80 per cent of the firms expect such results within two to three years. These findings are consistent with the view that there is a strong need for SMEs to engage in external collaborations targeted at short-term benefits due to their deficient R&D resources, as demonstrated in Figure 2. Although it is often thought that UIC activities are conducted utilizing scientific knowledge created by universities which pursue fundamental research, and that these activities typically focus on long run benefits, these findings indicate that this is not necessarily the case. Even for large enterprises with more than 2,000 employees, close to 40 per cent expect commercialization in two to three years, and most enterprises expect results from R&D collaborations within five years.

In order to examine further the attitude of firms toward UIC activities depending on firm size in detail, Figure 4 reports the obstacles that the firm encounters in UIC activities. For firms with fewer than 20 employees, 'insufficient funds' is by far the

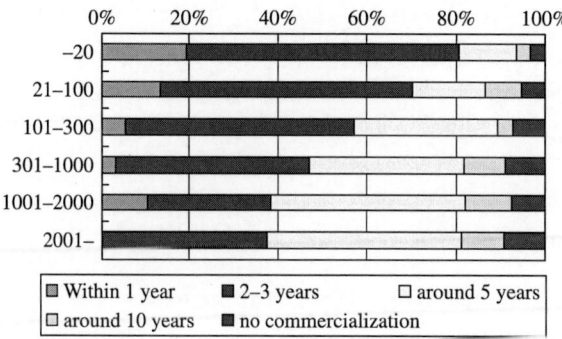

Figure 3. Timing of commercialisation

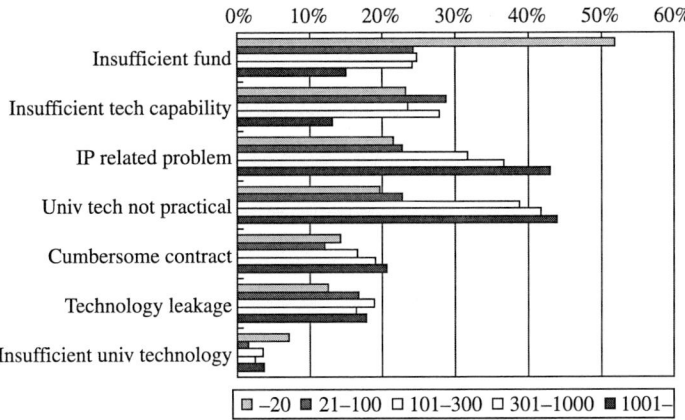

Figure 4. Obstacles for UICs by firm size

biggest obstacle. In addition, for small firms, the proportion of 'insufficient technological capability' is relatively high, indicating the essential role played by the firm's absorptive capacity when engaging in R&D collaborations. For large enterprises the proportions of 'university technology not practical' and 'IP related problems' are high. An interpretation of the former obstacle may be that although larger corporations are more inclined to pursue long-term joint R&D, they face a gap between their initial plans and reality at the stage of commercialization. On the other hand, it is interesting to observe that for small firms, the time span of R&D is shorter and their recognition of UIC activities as being 'impractical' is not as pronounced compared with large enterprises. In addition, the obstacle of 'IP related problems' may be interpreted as an indication of firms' sensitivity towards the movement of universities in strengthening their intellectual property management, exemplified by a series of university reforms such as the inauguration of Technology Licensing Organizations (TLOs) and the reform of the national university system.

Finally, it is worth noting that a rather high proportion of large enterprises have pointed to 'cumbersome contracts' as an obstacle. Presumably, contractual issues are serious problems faced by small and medium-sized firms, due to lack of staff for legal issues. As Motohashi (2005a) pointed out, the reason behind this might express differences between small and large enterprises in their approaches towards contracting. While large enterprises require lucid contracts with regard to the terms and the results of the collaboration, small firms can conclude contracts smoothly since the person who is responsible for the collaboration typically negotiates directly with university professors.[6] In addition, and as indicated earlier, small firms tend to have a more concrete image of the outcomes of the collaboration, and this may be a factor contributing to the relatively smooth signing of the contract by small firms.

R&D Environment Change and its Impact on R&D Collaborations

As mentioned already in this discussion, R&D collaborations exist in various forms, and there are certain characteristic features with respect to firm size.

This section discusses the changing nature of firms' R&D environment and its relation with external collaborations. Figure 5 shows the results of the items that the firm recognizes as being important factors in proceeding R&D.

As an overall trend, the proportions of 'meet market needs' and 'shorten lead times' are high. These results reveal the importance placed by firms on grasping market needs swiftly and pursuing R&D that reflects these needs. In addition, the proportions of 'focus R&D themes' and 'explore new R&D themes' are also high. At first glance, these results may look paradoxical. However, they may in fact demonstrate the firms' continual process of focusing and adjusting R&D efforts contingent on the changing needs of the market. On the other hand, the proportions of 'shifting to applied R&D' and 'upgrading basic R&D skills' both remained low. These results indicate that 'meet market needs' does not necessarily imply shifts to an application oriented research. However, they may in fact reveal the firms' current status of not being able to spare resources to improve basic research.

Looking at the trends in the different groups based on firm size, for small firms with fewer than 20 employees, the proportion of 'commercialize own technology seeds' is relatively high. This movement counters in direction with 'meet market needs', but simultaneously indicates the challenge faced by small-sized technology based firms to commercialize products using their own technology. For medium-sized firms – firms with between 21 and 300 employees – the proportions of 'meet market needs' and 'explore new R&D themes' are relatively high compared with other size groups. As the firm matures and reaches a certain size, it needs continually to supply newly developed products to the market. Finally, as for large enterprises – and especially the larger ones – the proportion of 'focus R&D themes' is relatively high compared with other size groups. Since the proportion

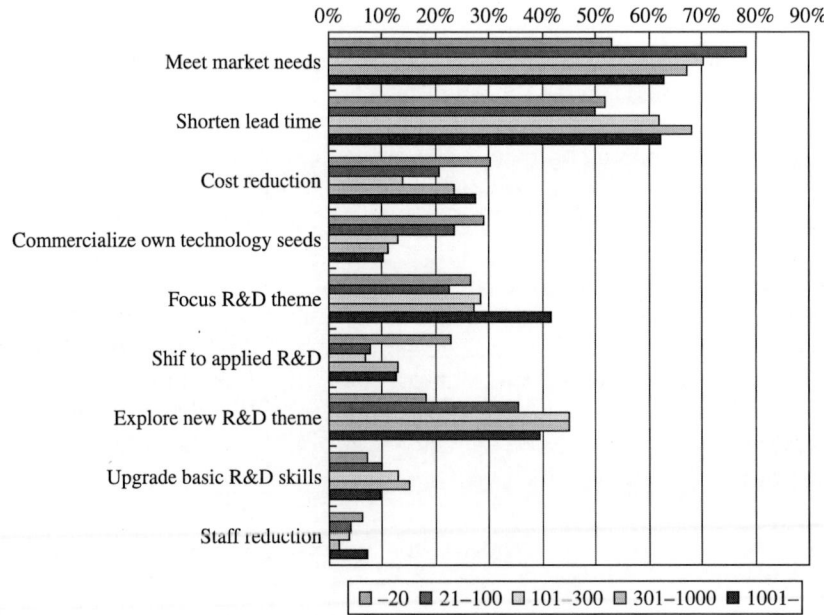

Figure 5. Important factors for R&D strategy

of 'discovery of new themes' is not low, this may be a reflection of firms' continual adjustment of R&D contingent on market needs, as previously indicated.

In order to investigate the relation between these items that the firms indicate as being important for R&D and external collaborations, a qualitative regression analysis is performed, using engagement in external collaboration as a dependent variable for each type of partner: large enterprises, SMEs and universities. The nine variables related to the environment of R&D listed in Table 1 are used as independent variables, and the logarithm of R&D budget, the logarithm of firm age, and the 35 industry dummies are included also as control variables. In order to illuminate the differences with respect to firm size, the whole sample is divided into SMEs with less than 300 employees and large enterprises with more than 301 employees, and a regression is conducted individually on both of these groups. Table 1 shows the signs of the coefficients of nine variables related to R&D environment, and only coefficients with statistical significance at the 10% level are displayed.

First, as for collaborations with large enterprises, SMEs that place importance on 'explore new R&D themes' and 'commercialize own technology seeds' are positively related. When small firms, especially start-up firms reach the stage of commercializing products using their own technology, collaborations with large enterprises that possess manufacturing facilities and marketing skills become vital. Many of the large enterprises collaborating with other large enterprises place importance on 'shorten lead times' in R&D. No outstanding pattern is found for collaborations with SMEs and start-up firms. SMEs that place importance on 'upgrading basic R&D skills' and large enterprises that place importance on 'focus R&D themes' exhibit a negative relation.

Finally, in respect of collaborations with universities small firms that place importance on 'cost reduction', 'shifting to applied R&D', and 'commercialize own technology seeds' show a positive relation. The theme of UIC activities

Table 1. R&D collaboration and R&D environment

	Large firms		SME and start-ups		Universities	
	SME	LF	SME	LF	SME	LF
Shorten lead time		++				++
Focus R&D themes				−−		++
Cost reduction					+	
Explore new R&D theme						
Shift to applied R&D	+++					++
Upgrade basic R&D skills				−	++	++
Meet market needs						
Commercialize tech seeds	++				++	+

Note: +++: positive coefficient and statistically significant at 1%
++: positive coefficient and statistically significant at 5%
+: positive coefficient and statistically significant at 10%
−−: negative coefficient and statistically significant at 5%
−: negative coefficient and statistically significant at 10%

probably centres around fundamental research, revealing the firms' strategy of pursuing fundamental research with universities and focusing in-house R&D on applied areas such as the commercialization of products using their own technology. In addition to these items, large enterprises show positive relations in areas such as 'shorten lead times', 'focus R&D themes', and 'explore new R&D themes'. As is the case with small firms, large enterprises also conduct fundamental research through UIC activities, increase the efficacy of in-house R&D process, and place importance on research that has close ties to the market.

Management of Firm Boundaries in Respect of R&D

As the business environment surrounding R&D changes vigorously, the R&D processes of firms are also changing. In order to proceed with R&D collaborations, it is necessary to clarify the relation of in-house R&D and such collaborations. In other words, the way a firm manages the boundaries of its R&D is becoming a decisive factor that affects the firm's performance in innovation activities (Nakamura & Odagiri, 2003). Finally, some survey results are presented of the way firms decide on whether to collaborate or conduct in-house R&D contingent on the nature of the R&D activity. Figure 6 shows the results of a survey asking whether the firm often conducts in-house R&D or pursues external collaborations by type of R&D content. For the latter case, the results are further broken down by the type of collaborative partner.

First, and as an overall trend, 'commercialization activities' and R&D for 'core technology' are often conducted in-house, while 'basic science' and 'technology frontier projects' are often conducted by external collaborations with universities.

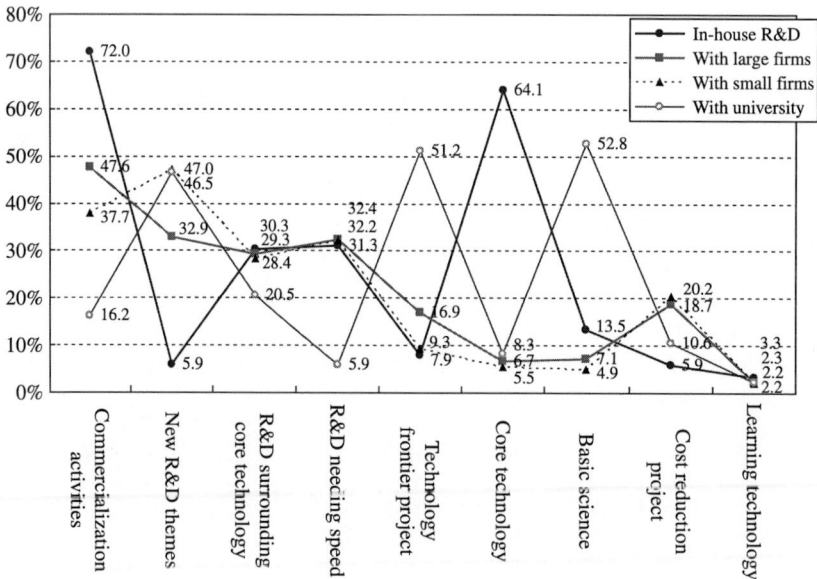

Figure 6. Collaborative R&D vs. in-house R&D

The proportion of firms conducting basic science in-house is around 10 to 20 per cent. The results indicate that, although most firms realize the importance of R&D in fundamental areas, in-house resources are focused on R&D for products close to commercialization, and fundamental research is left to R&D collaborations with universities. Although it is natural for a high proportion of firms to rely on external collaborations in new areas of R&D due to a lack of human resources and pertinent facilities, it is worth pointing out that a high proportion of these firms have chosen SMEs and start-up firms as their partners as well as universities. This trend is especially pronounced for large enterprises, indicating the increasingly important role played by NTBFs in the Japanese national innovation system.

UIC Activities and Research Productivity

Framework

In previous sections we illustrated trends in the changing nature of R&D environment and R&D collaborations. In this section, we investigate the relation between these activities and research productivity. R&D activities can be characterized as a process in which inputs – such as R&D investments and UICs – produce outputs, ranging from new products to production technologies. In order to analyse R&D productivity, it is essential to capture the output resulting from the input activities. In this current analysis, we use the number of patent applications by a firm as the output.

There are various problems with using patents as an output for innovations. For example, not necessarily all types of innovations are protected by patents: some may be kept as trade secrets, which is thought to be especially the case for process innovations such as manufacturing technologies. Also, appropriability conditions vary by industry and technology fields, and patent protection is particularly strong in the pharmaceutical industry (Cohen *et al.*, 2002a). A further problem is imposed by the great disparity in the quality of patents: using a simple aggregation of the number of patents does not accurately portray the output of innovation activities.[7]

Nevertheless, patent data is valuable and accessible data and indicates the R&D results of firms. Essentially it is the only index that covers different industries, and is widely used in quantitative analyses for innovation activities (Griliches, 1990). Therefore, the number of patents is used as the independent variable, and the following listed variables are used as the dependent variables in order to estimate the knowledge production function taking into account the effect of UICs:

- lrd: amount of R&D investments (in logarithm)
- lemp: firm size (logarithm of the number of employees)
- cord: dummy variable for R&D collaborations
- lage: firm age (year: logarithm)
- lage2: logarithm of firm age squared
- univ: dummy variable for UICs 5 years ago (in 1998)
- 35 industry dummies

The independent variables used are the number of patent applications in 2001 and the number of patents held at the time the survey was conducted

(February 2003). When using patent data as the independent variable, it is more appropriate to use the weighted average of the time-series R&D investment with an appropriate lag structure. However, since serial correlation is ordinarily strong in R&D data, using data from the same time period does not cause serious biasing, as shown in the case of the USA (Hall & Ziedonis, 2001). The coefficient of lrd is fairly stable in each of the cross sectional analyses at different time periods, indicating the presence of high serial correlation, hinting that the problem of using R&D data is minute as was reported in the case for the USA.

Results

Table 2 reports the results of the analysis using the number of patent applications in 2001 as the independent variable. It is found that the analysis using the number of patents held at the time of the survey shows similar results. First, Table 2 shows the results from the whole sample of the basic model with and without firm age (Model (2) and Model (1), respectively). Next the whole sample was divided into three groups based on age group, and the results for each of the groups are reported in Models (3) to (5).

First, in Model (1), the coefficient related to UICs in 1998 is positive and statistically significant, implying that engagement in UICs has a positive impact on research productivity. This coefficient does not change much in the model, including for firm age. For firm age a statistically significant result was not observed for the analysis using only lage, and an additional term was included – lage squared – and a statistically significant result was obtained for both of these coefficients. As the coefficients of these terms indicate, R&D productivity decreases with respect to the age of the firm, but after a certain threshold, increases once again, exhibiting a U-shaped pattern.

In order to delve into the effects of firm age and UIC, Model (3) to Model (5) show the results by firm age group. The firms in Model (3) are the oldest group, followed by those in Model (4) and Model (5). Statistically significant coefficients are found only in Model (5). This implies that young, relatively small firms have increased research productivity through UIC activities. As for the other group of firms, the result for Model (3) reports a positive coefficient although it is statistically insignificant, while Model (4) shows a negative coefficient. These results are consistent with the findings of Motohashi (2005a) which uses data collected in 2003 in RIETI's UIC Survey. Motohashi (2005a) interprets the results in the following way:

- Since SMEs are engaged in UIC activities that are closer to the final product stage such as the development of new products, these activities are likely to be more directly linked to the results of the development.
- As a background to the previous point, since SMEs are a constraint on both financial and human management resources, they have more need to engage in UIC activities within a shorter time scope than large enterprises.
- SMEs face a more stringent constraint on management resources than large enterprises, implying a greater risk in engaging in UIC activities. Therefore, the firms that were able to overcome these risks may be receiving a greater return.

Table 2. University industry collaboration and research productivity

	all	all	−1950	1951−70	1971−
	(1)	(2)	(4)	(5)	(6)
lrd	0.276 (7.81)**	0.260 ((7.19)**	0.434 (5.61)**	0.183 (3.05)**	0.190 (2.29)*
lemp	0.250 (6.08)**	0.246 (5.41)**	0.397 (3.72)**	0.315 (3.30)**	1.131 (2.84)**
cord	−0.030 (0.23)	−0.056 (0.45)	−0.131 (0.53)	0.146 (0.67)	−0.169 (1.06)
univl	0.377 (3.21)**	0.355 (3.05)**	0.203 (0.95)	−0.077 (0.33)	0.348 (2.09)*
lage		−2.402 (4.81)**			
lage2		0.360 (4.86)**			
Constant	−1.683 (7.10)**	2.302 (2.57)*	−4.257 (8.51)**	−1.188 (2.83)**	0.439 (1.30)
Industry dummies	yes	yes	yes	yes	yes
Observations	450	438	168	134	136
R-squared	0.62	0.64	0.77	0.55	0.49

Note: Absolute value of t statistics in parentheses. *significant at 5%; **significant at 1%

The results of the previous study also report that small firms that engage in UIC activities estimate a shorter time span until commoditization, indicating their strong awareness of R&D resource constraints, and show consistency with points 1 and 2. In addition, comparing the R-Squareds of Models (4) to (6), it can be seen that the R-Squared becomes smaller for younger firms. This indicates that the group of younger firms has a greater variance in performance and the proportion of the part that is unexplained by the independent variables is larger. This indirectly supports the claim made in point 3 about the differences in the risks faced by different types of firms.

Analysis of UICs with Respect to the Content of the In-house R&D

The results of the analysis in the previous section indicate that young, small firms have increased research productivity by engaging in UICs, and even for the group with the oldest firms, the coefficient of UIC is positive, although it is not statistically significant. In order to increase the firm's productivity by UIC activities, firms are required to possess absorptive capacity to migrate the fundamental scientific knowledge created by universities to their in-house innovation process. In this sense, it may be that larger enterprises have an advantage in utilizing UIC activities more effectively. The results of Table 2 exhibit a U-shaped pattern, implying that the effects of UICs decrease with respect to firm age, but increase once again after a certain threshold age, since both the aforementioned advantage due to smaller size and the factor of absorptive capacity are moving in both directions.

Examining the U-shaped relation in more detail, one can take a closer look at the relation between the content of in-house R&D and UIC activities. As shown in Sections 2–5, it is found that 'R&D close to commoditization' is often conducted in-house, while activities such as 'basic science' and 'technology frontier projects' are often conducted by collaborating with universities. In-house R&D and collaborative R&D are supposed to be systematically managed within the firm, so that the differences in the scope of in-house R&D may be affecting the performance of UIC activities.

In order to examine this linkage between in-house and collaborative R&D, the model of the previous section is extended and a regression is performed on the following one:

$$\ln Patent2001 = \alpha_0 + \alpha_1 \ln RD + \alpha_2 \ln EMP + \gamma \bullet (\beta_1 \ln Age + \beta_2 \ln Age \wedge 2) + \varepsilon$$
$$\gamma = 1 + \gamma_1 univ + \gamma_2 univ \bullet RD\_own \tag{1}$$

In the previous section, a regression was performed for the case that $\gamma = 1$, and for the regression based on different groups, the model that only univ interacts with γ. Here, an interaction term of age and univ is included, along with an interaction term of the content of the in-house R&D (RD_own), to clarify these relations. The independent variable is the number of patent applications in 2001 and the other independent variables are identical to the variables in the previous section except for RD_own. RD_own is a dummy variable representing each type of content of the in-house R&D. In addition, since one is comparing the effects of the contents of in-house R&D with that of external collaborations such as UICs, only the firms that are engaged in some form of external collaborations are used. Table 3 reports the results of the regression.

Model (1) is the base model that excludes RD_own. In order to capture the U-shaped characteristic of the effects of UIC on research productivity with respect to firm age, lage and the square of lage are included as independent variables. The coefficient of the interaction term of lage and univ is negative, and the effect of UIC decreases as firm age increases, but the coefficient of lage2 and univ is positive, indicating that the effect rises once again at a certain threshold value. In Models (2) to (7), although each of the γ's differ depending on the content of in-house R&D, the term of firm age that γ interacts with all include the square of lage, incorporating the non-linearity with respect to firm age.

The six types of in-house R&D picked here are 'commercialization activities', 'basic science', 'R&D needing speed', 'R&D for core technology', 'R&D surrounding core technology' and 'new R&D theme'. First, the inclusion of these variables does not significantly change the coefficients of Model (1). From in-house R&D content variables, statistically significant coefficients can be found with 'basic science', 'R&D surrounding core technology' and 'new R&D themes'. As for 'basic', the coefficient of the interaction term with univ*lage is positive and the coefficient of the interaction term with univ*lage2 is negative. The results for 'R&D surrounding core technology' and 'new R&D themes' show opposite signs.

In interpreting 'basic science', the results indicate that the effects of firm age on the impact of UICs (interaction term of univ and lage) are affected by whether or not the firm conducts this type of in-house R&D, and for firms that conduct such in-house R&D: the older the firm, the greater the effects of UICs. However, since the coefficient of the interaction term with univ*lage2 is negative, this trend reaches its peak at a certain firm age at which point it starts to drop, exhibiting a reversed U-shaped pattern. 'R&D surrounding core technology' and 'new R&D themes' exhibit an opposite pattern to this, implying that for firms that are conducting these types of in-house R&D, the effects of UIC decrease as firm age increases.

The reason that the effects of UICs exhibit a U-shaped nature with respect to firm age is the difference in the focus of in-house R&D by each of the firms. Old, large enterprises that have strong infrastructures for research are able to conduct basic science research that is complementary to UICs. On the other hand, small firms face a severe resource constraint for research, and firms that are actively involved in new R&D themes show positive effects from UICs. These firms are thought to be involved in UIC activities that require a shorter time period for commercialization than larger enterprises; they are engaged in UICs with a clearer target in mind.

The Role of SMEs and Start-ups in Changing the Japanese National Innovation System

Hitherto the role and the effects of UICs within the context of Japanese firms' innovation processes have been discussed. Large enterprises played a pivotal role in the Japanese national innovation system, but recently, new technology based

Table 3. UIC, productivity and in-house R&D

	lpat2001 (1)	lpat2001 (2)	lpat2001 (3)	lpat2001 (4)	lpat2001 (5)	lpat2001 (6)	lpat2001 (7)
lrd	0.293 (6.97)**	0.294 (7.04)**	0.285 (6.84)**	0.295 (7.06)**	0.285 (6.76)**	0.300 (7.11)**	0.290 (6.94)**
lemp	0.221 (4.22)**	0.214 (4.09)**	0.220 (4.25)**	0.216 (4.13)**	0.214 (4.13)**	0.220 (4.22)**	0.226 (4.35)**
lage	−0.280 (0.80)	−0.313 (0.89)	−0.320 (0.91)	−0.320 (0.91)	−0.314 (0.90)	−0.251 (0.71)	−0.291 (0.83)
lage2	−0.008 (0.12)	0.001 (0.02)	0.003 (0.04)	0.001 (0.02)	0.002 (0.02)	−0.012 (0.16)	−0.003 (0.05)
univ*lage	0.193 (3.22)**	0.172 (2.43)*	0.213 (3.48)**	0.205 (3.25)**	0.120 (1.72) +	0.167 (2.74)**	0.178 (2.98)**
Commercialization*univ*lage		−0.076 (0.29)					
Commercialization*univ*lage2		0.032 (0.50)					
Basic science*univ*lage			0.766 (2.22)*				
Basic science*univ*lage2			−0.162 (2.00)*				
R&D needing speed*univ*lage				0.171 (0.58)			
R&D needing speed*univ*lage2				−0.056 (0.76)			
Core technology*univ*lage					−0.357 (1.39)		
Core technology*univ*lage2					0.106 (1.65)		
R&D surrounding core tech*univ*large						−0.550 (1.65)	
R&D surrounding core tech*univ*large2						0.139 (1.69) +	
New R&D theme*univ*lage							−1.27 (1.85) +
New R&D theme*univ*lage2							0.330 (1.87) +
Constant	−0.647 (1.20)	−0.621 (1.15)	−0.550 (1.02)	−0.612 (1.14)	−0.538 (0.99)	−0.751 (1.38)	−0.656 (1.22)
Observations	345	349	349	349	349	349	349
R-squared	0.64	0.65	0.65	0.65	0.65	0.65	0.65

Note: Absolute value of t statistics in parentheses. + significant at 10%; *significant at 5%; **significant at 1%

firms as well as start-ups have also started to engage in UICs, and have achieved successful results compared with large enterprises by focusing R&D themes. What impact does the rapid advance of these small firms and start-ups have on Japanese innovation systems? In this section, the role of SMEs as a partner of R&D collaborations of large firms are analysed.

First, the trends are reported in the content of collaboration between SMEs and start-ups grouped by firm age. As can be seen in Figure 7, the content of the collaborations with SMEs is wide-ranging. The types of content with high proportions are 'new R&D themes' and 'commercialization activities'. Firms with collaborations for 'new R&D themes' intend to widen their R&D scope, while firms with collaborations for 'commercialization activities' are seeking for faster innovation by using R&D supports from universities. In addition, the number of spin-offs from universities that base their technology on technologies developed by the university has been increasing recently, and most types of collaborations with these firms probably fall under the 'new R&D themes' category (see study by Debroux in this current collection).

In addition, a relatively high number of firms listed the contents of 'R&D needing speed', 'R&D surrounding core technology', and 'cost reduction projects'. As the R&D environment changes vigorously, firms may be choosing from a pool of different types of SMEs and start-up firms in order to meet their wide range of needs. As for the relation between the groups divided by firm size, the group with the relatively younger firms has a high proportion of 'commercialization activities', while the group of older enterprises, which are larger and more established, possess a high proportion of 'new R&D themes'. This indicates that large enterprises are utilizing SMEs and start-ups to exploit new areas related to the development of new technologies.

In order to further examine the determinants of collaborations with small firms and start-ups, a regression is performed as to whether or not the firm engaged in such collaboration as the dependent variable and lage, lrd, 7 types of RD_own used in the previous section and industry dummies as independent variables. A probit model is used for the analysis and the results are shown in Table 4.

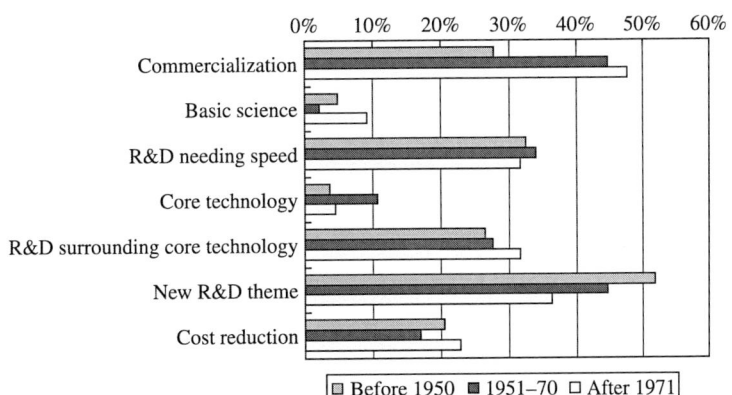

Figure 7. R&D collaboration contents with SMEs and start-ups

Table 4. Determinants of collaboration with SMEs and startups

	Contents of R&D collaboration with SME and start-ups						
	5 yrs ago (1)	Present (2)	Com (3)	Speed (4)	Surr (5)	New (6)	Cost (7)
lage	0.048 (0.26)	0.281 (2.58)*	0.096 (0.76)	−0.121 (1.03)	0.068 (0.53)	0.120 (1.01)	−0.065 (0.47)
Lrd	0.144 (2.76)**	0.035 (0.80)	−0.120 (2.32)*	0.092 (1.80) +	−0.047 (0.87)	0.125 (2.53)*	0.074 (1.33)
Commercialization (Com)	0.259 (1.09)	−0.003 (0.01)	0.233 (0.93)	−0.517 (2.05)*	−0.067 (0.27)	−0.038 (0.16)	0.267 (0.88)
Basic science (Basic)	0.590 (1.80)	1.172 (3.57)**	0.587 (1.91) +	0.151 (0.46)	0.188 (0.53)	0.019 (0.06)	−0.081 (0.20)
R&D needing speed (Speed)	0.017 (0.08)	0.138 (0.67)	−0.363 (1.42)	−0.097 (0.39)	0.345 (1.43)	0.065 (0.29)	0.004 (0.01)
Core technology (Core)	0.236 (1.03)	0.307 (1.49)	0.409 (1.65) +	0.627 (2.25)*	0.471 (1.82) +	0.379 (1.63)	0.050 (0.17)
R&D surrounding core tech (Surr)	0.003 (0.01)	−0.186 (0.88)	−0.193 (0.77)	−0.187 (0.71)	−0.030 (0.12)	−0.190 (0.82)	0.423 (1.49)
New R&D theme (New)	0.008 (0.02)	0.056 (0.14)	−0.284 (0.58)	0.448 (0.95)	0.751 (1.84) +	0.105 (0.24)	0.346 (0.68)
Cost reduction (Cost)	0.209 (0.51)	0.400 (1.00)	1.183 (2.82)**	−0.820 (1.35)	0.346 (0.76)	−1.271 (1.97)*	0.375 (0.76)
Constant	−1.795 (1.59)	−1.141 (1.14)	0.303 (0.30)	−0.533 (0.48)	−0.229 (0.19)	−2.310 (2.01)*	−1.176(0.90)
Observations	205	232	218	231	219	228	198

Note: Absolute value of z statistics in parentheses. + significant at 10%; *significant at 5%; **significant at 1%

RD_own is a variable available only for firms that are involved in external collaborations as was the case in the previous regression, thus the sample of the regression includes firms that are engaged in some form of external collaboration.

Model (1) and Model (2) indicate the characteristics of firms that conducted collaborations with small firms and start-ups FIVE years ago and the characteristics of firms that are presently involved in such collaborations. Both of the analyses reveal that large enterprises with a relatively older age and larger R&D budgets are more actively involved in collaborations with SMEs. As regards to the variable related to the content of the in-house R&D, 'basic science' has a positive effect and this effect is especially marked more recently. This implies that firms conducting fundamental research in-house have been collaborating with SMEs. This trend especially has become more pronounced, indicating the increasingly important role of SMEs and start-ups in the innovation process of large enterprises.

Models (3) to (7) show the results of a similar regression based on the content of the collaboration with SMEs and start-up firms. The sign of coefficients of lrd depends on the type of R&D collaboration content. For example, small firms with a relatively small R&D budget collaborate more with SMEs in 'commercialization activities', while large enterprises that possess a large budget collaborate more in 'R&D needing speed' or 'new R&D themes'. As regards to the variable related to the content of in-house R&D, firms that focus on 'core technology' actively collaborate with SMEs in 'commercialization activities', 'R&D needing speed', and 'R&D surrounding core technology'. Firms are focusing their in-house R&D on core technologies and obtaining the necessary services to speed up their R&D process by external collaborations, and these results may be indicating the advances in selection and concentration for in-house R&D.

Conclusions and Implications for the Japanese Innovation System

It has been found that young, relatively small sized, new technology based firms have managed to sustain high productivity in R&D by collaborating with universities – an insight developed elsewhere in this current collection. Absorptive capacity is critical in order for firms to make use of the fundamental research results possessed by universities for their in-house R&D process, but NTBFs have had successful results in UICs by focusing on types of activities that are closely related to commercialization and have concrete objectives. Since these firms do not have sufficient funds and human resources compared with large enterprises, they have a bigger incentive to collaborate actively in R&D in areas close to commercialization, such as new product developments. However, engagement in UICs may be seen as an investment with great risks for small firms that only have scarce resources. Small firms which succeed in UIC activities may be rewarded by larger returns than large firms, because UIC projects are relatively risky for small firms with a lack of managerial resources.

These NTBFs that conduct external collaborations take risks, have a great potential for growth, and thus the potential for becoming a driving force in the reform of the Japanese innovation system. An increasing number of firms are exploiting R&D collaborations with SMEs and venture start-ups. There is an emerging trend for large enterprises focusing on in-house R&D in fundamental research to engage in such collaborations. Corporations that conduct business globally may also have global alliances in order to gain access to essential technologies; ideally, they would prefer to have allies within geographical proximity. Therefore, the emergence of SMEs and start-ups that possess different types of technologies is also crucial from the standpoint of enhancing innovation performance of large enterprises.

The rapid advance of NTBFs is vital from the perspective of the Japanese innovation system as a whole. Owing to enduring and systematic impediments in the Japanese innovation system such as the rigidity of the labour market and the underdevelopment of the technology market, active involvement in R&D collaborations between enterprises did not flourish. Consequently, large enterprises played the pivotal role in the innovation process empowered by their rich resources in R&D. However, due to the reliance on in-house R&D, there is a possibility of Japanese firms being left behind in competition for innovation activities, and especially in areas where technological process is rapid, as exemplified by the information technology (IT) revolution (Ando & Motohashi, 2002). Moreover, in the pharmaceuticals R&D process – changing rapidly due to advances in biotechnology – industry collaborations are crucial between universities and other institutions offering expert scientific knowledge in fields such as genetic engineering (Motohashi, 2005b). In the high technology industries such as IT and biotechnology, the network-based system of R&D collaborations has started to gain competitive advantage.

Because NTBFs lack R&D resources, they have a strong incentive to engage in UICs that lead to immediate results such as the development of a new product, overcoming the systematic impediments of networking. Universities also have an incentive to collaborate with NTBFs, since joint projects with NTBFs tend to fulfil their desire of creating pragmatic products based on the results of fundamental research, in contrast to R&D collaborations with large enterprises that tend to prefer fundamental research projects. The active involvement of NTBFs in UICs – overcoming the aforementioned systematic impediments – has a great potential to reform the system of innovation as a whole, diminishing its rigidity. Therefore, active engagement between NTBFs in UICs can have a catalytic effect in reforming Japanese innovation systems. Such activities are critical from a policy perspective and should be promoted further.

In order to stimulate start-up firms and NTBFs, systematic reforms for transforming Japanese innovation systems towards a network-oriented system are vital, as is direct assistance through taxation policies and subsidies. As suggested above, a number of firms listed 'intellectual property related problem' as an obstacle. Many of the enterprises expect licensing from foreign universities and enterprises to increase in the future, indicating the firms' increasing awareness of international disputes regarding patents. An active technology market with a stable system of intellectual property is a necessary condition to foster high-technology start-up firms that lack the managerial resources to provide input in areas such as manufacturing and marketing (Hall & Ziedonis, 2001). In addition, the reforms of national universities and national research institutes have enabled these organizations to collaborate with enterprises at their own will. The system for stimulating NTBFs in UIC activities and facilitating university spin-offs that use the technologies of universities is improving, as demonstrated in other studies presented in this current collection. In addition, restructuring the capital market is a vital task in order to facilitate a seamless supply of risk money and fostering venture capital.

As a final note, in order to establish a societal system replete with innovative activities through R&D collaborations between various parties, the mobility of human resources becomes crucial. The rigid system of personnel management that is common in some enterprises or research institutes may become an obstacle for realizing effective collaborations amongst firms or between firms and universities. Increasing the mobility of human resources offers various new career opportunities for researchers. This may result in attracting more stars, thus stimulating the system of innovation as a whole. Increasing the mobility of human resources is an intricate problem due to the inherent and enduring complexity of issues such as organizational constraints, the employment system, and firm specific customs: for example, as expressed in the traditional legal constraints discussed subsequently by Taplin in this collection. This enduring complexity remains a pivotal issue, and one that needs resolution in order to establish a new type of Japanese innovation system based on active involvements of start-up firms and NTBFs.

Acknowledgements

The author would like to thank Masaru Yoshitomi, Yuji Hosotani and Toshihiro Kodama at RIETI for their helpful comments. The views expressed in the study are those of the author's and do not reflect the views of his organization.

Notes

[1] This contribution is based on RIETI's research project on 'R&D Collaborations and Japan's innovation system', and the Japanese version of this paper is published as RIETI discussion paper 2005-J-002.

[2] The survey results of METI (2003b) collected for the White Paper on Small and Medium-Sized Enterprises in Japan are an exception that does not suffer from sampling problems.

[3] For detailed results of RIETI's R&D External Collaboration Survey, refer to RIETI (2004).

[4] This U-shape pattern does not result from biases from the distribution of firms based on industrial classifications or types of technology (for example, the majority of small firms are biotechnology or IT start-ups), but instead reflects a common characteristic observed across industries. For a closer examination of the sample size grouped by firm sizes and industrial classification, refer to RIETI (2004).

[5] The exceptions were medical-related start-ups (founded within five years and with the number of employees less than 500), which actively took part in industry-academia collaborations.

[6] This point is also noted in the results of an interview conducted with university staff who have been involved in UIC activities (RIETI, 2001).

[7] There is a possibility that the younger the firm, the lower the average patent quality and the higher the propensity to apply for patents. Although analyses have proceeded using patent citation data in the USA (Hall, Jaffe and Trajtenberg (2001)), a similar database has not yet been developed in Japan.

References

Acs, Z., Audretsch, D. & Feldman, M. (1994) R&D spillover and recipient firm size, *Review of Economic and Statistics*, 76, pp. 336–340.

Ando, H. & Motohashi, K. (2002) Japanese economy, the structure of competitiveness: modularization strategy challenges, in: *The Age of Speed* (Tokyo: Nihon Keizai Shimbun) (in Japanese).

Cohen, W. M. & Levinthal, D. A. (1990) Absorptive capacity: a new perspective on learning and innovation, *Administrative Science Quarterly*, 35, pp. 128–152.

Cohen, W., Goto, A., Nagata, A., Nelson, R. & Walsh, J. (2002a) R&D spillovers, patents and the incentives to innovate in Japan and the United States, *Research Policy*, 31, pp. 1349–1367.

Cohen, W., Nelson, R. & Walsh, J. (2002b) Links and impacts: the influence of public research on industrial R&D, *Management Science*, 48(1), pp. 1–23.

Griliches, Z. (1990) Patent statistics as economic indicators: a survey, *Journal of Economic Literature*, 28 (December), pp. 1661–1707.

Hall, B., Jaffe, A. & Trajtenberg, M. (2001) The NBER patent citation data file: lessons, insights and methodological tools, NBER Working Paper Series 8498.

Hall, B. & Ziedonis, R. (2001) An empirical study of patenting in the US semiconductor industry, 1979–1995, *Rand Journal of Economics*, 32(1), pp. 101–128.

JASME (Japan Finance Corporate of Small and Medium Size Enterprises) (2002) Current state and obstacles of university industry collaborations by SMEs, JASME Report No. 2001-04, February (in Japanese).

Kodama, T. (2003) Technological innovations and the cluster formation of TAMA firms – based on a survey result, RIETI Discussion Paper Series 03-P-004, February (in Japanese).

METI (Ministry of Economy, Trade and Industry) (2001) Report on Japanese Innovation System Research, METI, July (in Japanese).

METI (2003a) On the collaboration of enterprises with universities and research institutes, General document, METI, 25 April (in Japanese).

METI (2003b) White Paper on small and medium-sized enterprises in Japan, METI, May (in Japanese).

Mitsubishi Research Institute (2002) Report of the 2001 Survey on SMEs regarding Management innovations, The 2001 Report on SMEs for the Small and Medium Enterprise Agency, March (in Japanese).

Motohashi, K. (2001) *The Current State and Problems Faced by Japanese Innovation System* (Tokyo: RIETI).

Motohashi, K. (2004, June 25) The current state of UICs and the importance of NTBFs – implications for Japanese innovation system, *Kaihatsu Gijutsu*, 10, pp. 1–15 (in Japanese).

Motohashi, K. (2005a) Economic analysis of university-industry collaborations: the role of new technology based firms in Japanese national innovation reform, *Research Policy*, 34(5), pp. 583–594.

Motohashi, K. (2005b) The changing autarky: pharmaceutical R&D process, causes and consequences of growing R&D collaboration in japanese firms, *International Journal of Technology Management*, 39 (1/2), pp. 49–71.

Nakamura, K. & Odagiri, H. (2003) Determinants of R&D boundaries of the firm: an empirical study of commissioned R&D, joint R&D, and licensing with Japanese company data, National Institute of Science and Technology Policy (NISTEP) Discussion Paper No. 32.

RIETI (Research Institute for Economy, Trade and Industry) (2001) Report on the Japanese National Innovation System Research, July 2 (in Japanese).

RIETI (2003) Report of University Industry Collaboration Survey, May (in Japanese)

RIETI (2004) The 2003 Report of the R&D External Collaboration Survey, June (in Japanese).

Wen, J. & Kobayashi, S. (2001) Exploring collaborative R&D network: some new evidence in Japan, *Research Policy*, 30, pp. 1309–1319.

Japanese Intellectual Property and Employee Rights to Compensation

RUTH TAPLIN

Introduction

Historically, the relationship of the Japanese to legal matters has tended to be an ambivalent one; borne out of necessity but with a sense of shame, litigation was prohibited between vassals and overlords. Japan has had to overcome this societal constraint to adopt western patent law, to enable it to industrialize and become a global force.

It has always been the case that cyclical periods of economic crisis increased patent activity, which was then followed by prosperity. This cycle has repeated itself since the sixteenth century when Japan first began to industrialize with the Mitsui Zaibatsu (a large traditional family owned company) taking the lead through its incipient role as bankers to the nobles and as major traders with China and beyond.

In the sixteenth century, the Japanese became the world's leading manufacturers of firearms in the world; they had perfected such manufacturing from the Portuguese. However, in the repressive Edo Shogunate lasting 250 years the Japanese fell behind the West in technological development. In 1854, Japan emerged from its isolation and looked to patenting to invent its way out of economic stagnation and technological underdevelopment. The first patent ordinance was passed on 18 April 1885, but as it did not recognize the rights of foreigners, a new Patent Law was introduced in 1899 that did so, and thereafter, Japan joined the Paris Convention. After the Second World War, Japan again in crisis invented its way into a prosperity admired and envied by western countries with such notable inventors as Konosuke Matsushita, the founder of the Matsushita group, who set up his company based on the proprietary rights to a new electronic socket. In the development of his global electronics business, he acquired over 100 patents.

However, after the spectacular success that the Japanese experienced during the 1960s when they became lauded for their inventiveness, total quality management procedures based on models proposed by Deming and general business success fortunes declined, bringing on economic stagnation. Invention stagnated, with universities giving little incentive to encourage inventions of their staff to be patented and licensed. The relationship between the university inventors and companies was poor and invention internal to companies was not properly rewarded. Entrepreneurship was not encouraged, and coupled with the cultural idea that 'the nail that sticks up should be hammered down'; this meant that some of the most inventive Japanese were lured by the rewards offered by the United States. The patent system had not changed substantively since 1960 and was becoming hopelessly out of date in a world where intangible assets were beginning to comprise 70 per cent of most major companies' assets. The wake up call came about four years ago when the reality of the world's second largest economy becoming globally uncompetitive forced the Japanese government to act.[1] This paper looks at some of the major changes spurred on by this realization and analyses the implications for Japan of such changes.

Research Proposition

Overall, this study advances the following proposition:

- Under the Koizumi Government (2001 to 2006) the promotion of intellectual property (IP) propelled the Japanese economy out of its malaise and gave rise to inventiveness and a proliferation of innovative small and medium sized enterprises (SMEs) – companies not seen since the 1960s. The changes brought throughout Japanese society by the IP drive have been multilayered and have laid the foundations of a new strengthened economy based on innovation.

This essay develops an historical and economic perspective on how innovation linked to IP has provided a new impetus for inventiveness in Japan. It suggests also how this impetus has implications for the process of reforming the traditional structures and cultures that make up Japan's national innovation system (NIS).

Context: TLO Laws, Privatization of Universities, New IP Divisions

Technology licensing organizations (TLOs) are effective vehicles to create closer ties between universities and industry. The TLO law or 'Promoting

University-Industry Technology Transfer' has ushered in a new relationship between inventors, universities and industry. Japan initially followed the strategies of the USA and then adapted these TLO policies to the Japanese case.

As Terry Young, former President of the Association of University Technology Managers (AUTM) and manager of technology transfer at Texas A&M University notes, technology transfer is needed to reward, retain and recruit faculty members. To understand the process of commercialization, which allows the assessment of the value of IP, it is important to encourage a better and closer relationship between universities and industry (Young, 2004).

As illustrated already by Ibata-Arens in this current collection, TLOs in the USA serve as engines for economic development by spinning out new companies and creating new employment opportunities. TLOs manage a process of commercializing and transferring research results for public good and benefit.

By 2006, Japan had adapted the TLO idea to 34 mainly University centres whereby TLOs which are approved can use national university properties without any cost (the centres were privatized in April 2004). The role of TLOs has expanded further from solely technology-transfer functions to centres that assist university start-ups. Professor Akio Nishizawa, Deputy Director of the New Hatchery Centre at Tohoku University, points to a development unique to Japan; that of intra-university IP Divisions. The latter exist solely to give advice and support to those in the university who wish to bring their inventions to fruition. The Ministry of Education, Sport, Science and Technology (MEXT) has awarded such IP Divisions to 34 Japanese universities (Nishizawa, 2006).

Professor Nishizawa points to further structural changes that need to be implemented in the Japanese economy which are all inextricably linked to intellectual property. These changes include moving from a follow-through to a break-through system, from centralized to decentralized, from government (bureaucracy)-led to market-oriented, changing from big business with scale of economy to entrepreneurial ventures with scope of economy and from capital labour intensive to knowledge intensive. The success of the Japanese TLOs may be seen in the rise in number of university start-up ventures in Japan, which saw a rise from 98 in 1995 to 1,112 in 2004. This has met the target for spin-offs but Professor Nishizawa notes that the performance of university start-ups is not good enough to boost the regional economy significantly by attracting high technology industry. Unlike the USA there is a lack of risk money and highly talented people.

Although the technology transfer system between university and industry and university structure has changed, there has been little change in Japanese financial and labour policy to match. Venture Capital activities remain miniscule compared to those in the USA and Europe.

Employees' Rights to Compensation

In Japan, employees' rights to compensation have not changed substantially in law since the Patent Law of 1960. Employees of both companies and businesses have been reluctant to invent or licence their inventions because of the poor relations between business and universities, the fact that universities could not own their own IP, and in general less than satisfactory terms of remuneration. This has resulted in inventions stagnating and emigration to the United States. A feature equally worrying for Japanese business and economy in addition to emigration were the high settlements being granted by the Tokyo High Court in 2004, with the Nakamura Shuji vs. Nichia becoming a landmark decision by the then presiding Judge Shitara. Since then there have been modifications to the law that balance the rights of inventors with that of employers.

The first attempt to deal with unrealized inventions has been by Dr Kimio Ishimaru, Director of the National Institute of Advanced Industrial Science and Technology (AIST). AIST was instituted on 1 April 2001 and was a new research organization comprising 15 research institutes that were formerly under the Agency of Industrial Science and Technology in the Ministry of International Trade and Industry and the Weights and Measures Training Institute. The restructured AIST is Japan's largest public research organization with many research facilities and around 6,000 employees in total.

According to Dr Ishimaru, a recent Director, the portfolio consisted of 10,000 research projects in various states of completion from all affiliated institutes. Since 2001, over 150 licensing agreements have been made and are concluding at 30 agreements a year, increasing by at least ten a year. Many of the inventions are fundamental ones – for example, a core liquid crystal display (LCD) patent belonging to AIST, which is a leaner, less bureaucratic institution that is designed to maximize the advantages of an independent administrative body, and to ensure the autonomous functioning of the organization.

The more pressing problem has been, in the author's estimation, how to deal with the potential flood of compensation claims by current or former employees who do not feel they have been compensated properly. Success by others in obtaining adequate compensation and comparisons with colleagues in the United States and the United Kingdom can only lead to growing resentment on the part of the inventors. Japanese inventors have seen how well they may be compensated in the Nakamura Shuji case and when they travel abroad. Japan can ill afford mass migration of bright inventors to the USA and Europe.

The recent debate concerning compensation, according to Professor Katsuya Tamai, Head of CASTI, the number one TLO in Japan based at Tokyo University and a major expert on employees' rights to compensation, revolves around whether to follow the German method that is favoured by scholars or the American method that is seen as superior by business. Professor Tamai has noted that despite the rigour of the German system it has failed and is not appropriate for Japan. The American system is preferred but Professor Tamai argues that Japan is not ready for its complete adoption, as freedom of profession does not exist in Japan where salary scales are still linked to age and lifetime employment as well as company loyalty. He suggests as the safeguarding of employees' rights is paramount, in line with new moves in German law, incremental payments should be allowed that occur at different points of patent procedures, such as registration. With respect to losing good researchers and inventors to the United States, a US-type contract system should be adopted, with equal bargaining power for both employers and employees (Tamai, 2003).[4]

Professor Tamai appeared on NHK television in February 2004 explaining the significance of the award of 20 billion yen to inventor Shuji Nakamura by Nichia

Corporation. Nakamura was instrumental in the development of a light emitting diode (LED) semiconductor, which glows blue when electricity is passed through it.

This invention could net Nichia Corporation at least 120.8 billion yen in profits through its exclusive ownership rights up to October 2010. The Tokyo District Court calculated that Nichia should pay Nakamura half of the potential profit amounting to 60.43 billion yen, but since Nakamura had asked for 20 billion yen the court ordered Nichia to pay that amount as compensation for his invention. In January 2005, Nakamura settled with Nichia Corporation for 844 million yen (US$8 million) as the Tokyo High Court overturned the decision of the District Court that had initially awarded the inventor 20 billion yen (US$187 million). The Japanese Business Federation was relieved, as were other businesses around the world. The Tokyo District Court however, found that Mr Nakamura was solely responsible for the invention contrary to the statement issued by Nichia that the invention was a joint effort. Nichia had bought the invention for US$2,700,000 and had assessed in a negative light the potential value of the blue emitting diode, even ordering Mr Nakamura to cease research work on this tremendously important invention at one point. This was why the contribution to the employer was deemed to be only 50 per cent.

> It is very difficult to legislate for the lack of vision by institutions and employers. It also appears that without such a ruling by the Tokyo District Court inventors would have confirmed in their minds that they will never be remunerated properly in Japan and they would become de-motivated and then emigrate elsewhere.[2]

The Nakamura case is not an isolated one. The day before the Nakamura ruling, the Tokyo High Court ordered electronics company Hitachi to pay their ex-employee Seiji Yonezawa what was then a record 163 million yen (US$1.5 million) for his contribution to optical disc technology. Prior to this judgement Hitachi had compensated Yonezawa with only 2.3 million yen for early stage DVD technology. In a landmark case in 2004, *Olympus* v. *Tanaka*, the Supreme Court made it clear that inventors can sue their companies for larger shares of the profits resulting from a successful invention, irrespective of employee agreements and internal rules. This appeared to give the courts total discretion to allocate size of reward, which has been evidenced by the Yonezawa, Nakamura and many other subsequent cases. This has resulted in litigation over intellectual property rising to over 700 cases in 2003 from virtually none over a decade before, while there were 602 complaints filed with the Tokyo High Court concerning decisions made by the Patent Office, including 421 over patent rights. In Japan, a country traditionally adverse to litigation, this marks a significant change. An amendment to Article 35 of Japan's Patent Law was drafted to protect employers by making the courts respect an employment agreement or company rules on invention compensation unless such compensation is deemed unreasonable. In 2006, Section 4 of Article 35 was amended to:

> The amount of reasonable remuneration shall be decided by considering the amount of profit that the employer shall obtain from the invention and

burdens that accompany the inventions on the employers and the degree of the contribution by the employer, and the benefits given to the employee by the employer and in all other circumstances.

This provision is a counterbalance to the original Section 4 which allows employers redress as it ensures that contributions from the employers' side are taken into account in the assessment of worth of inventions. This will lead to other cases for compensation paying greater attention to the valuing of IP that can be written into initial employee agreements, but needs to be more accurately assessed.

Valuing IP in Japan

The valuing of IP is becoming an issue of increasing concern and importance. Taisuke Kato, General Manager of Toshiba IP Division Corporate Headquarters in Tokyo, noted how Toshiba understands the importance of placing a value on IP. In developed economies, the migration from manufacturing to services is paralleled by a shift in asset evaluation from physical assets to intangible assets. A well-known example of this according to Kato is the estimate that two-thirds of the market capital of listed companies in the United States can be accounted for by intangible assets, including IP.

While there is increasing discussion on how best to value IP, a unified methodology has yet to emerge. The market value method has its supporters, as it tries to consider not only present royalty streams but also possible future royalty income. However, there remain difficulties in evaluating certain IP assets, particularly future potential. Toshiba is now looking at the promotion of a system of relative evaluation that measures their strength in IP against that of their competitors. The consideration here is quantity versus quality. Quantity can easily be determined, while quality is more problematic, but possible to some extent.

Toshiba is a Japanese company that has integrated successful IP valuation into the three-tier incentive system that they use to encourage and reward innovation among their researchers and engineers. The first tier is transfer remuneration, under which an inventor can receive as much as 15,000 yen for transferring an idea to Toshiba. Business remuneration is paid – if and when that invention is used in product – and licensing remuneration is paid from any licensing income Toshiba might receive for IP. The other two incentives depend on the market and the extent to which the IP is utilized. Every year, the company arrives at a value for productive IP by calculating how much it contributed to actual sales and earned in royalty income (Taplin, 2004). This gives them the basis for open-ended business and licensing remuneration incentives but has not made them immune to their inventors challenging them to providing fair remuneration for inventions.

Changes to Patent Courts and Status of the Patent Attorney (Benrishi)

Japan's commitment to invention is profound. Prime Minister Koizumi launched the national strategy for invention and IP in a policy speech to the Diet in early 2002, resulting in March of that year in the formation of the Strategic Council on Intellectual Property. On 17 April 2003 a Day of Invention occurred to galvanize the nation into consensus-based action for the promotion of invention and IP. Along with all the sweeping reforms of Patent Law in general are other historic changes to the Patent Attorneys Law, not seen since the beginning of the Meiji Era, and the new introduction of Patent Courts in 2005 (*Japan Times*, 2004).

Sumiko Shimosaka, President of the Japan Patent Attorneys Association (JPAA), outlined a number of key areas for JPAA to concentrate on. These included the problems of counterfeiting and infringement of IP in Asia; participating more in international conferences and working together with more international IP organizations. Paramount is the training of patent lawyers to have the expertise to cope with the new complex reforms. The increase in patent attorneys obtaining the new expertise has already reached a milestone with 533 *benrishi* having passed an examination that complies with a recent law allowing patent attorneys to jointly and equally represent clients with attorneys at law (*bengoshi*). The fact that *benrishi* can now represent their clients equally changes the whole process of IP litigation. At a symposium Sumiko Shimosaka noted that *benrishi* are now set to assume an important role in creating, protecting and utilizing IP.[3]

In January/February 2004, the Strategic Council on Intellectual Property and the Office for Promotion of Justice Systems Reform advisory panel to the prime minister decided that an independent high court should be created by April 2005. This decision followed many months of acrimonious debate with some arguing for a new independent high court specializing in IP infringement cases that would make the litigation process easier for businesses and become a ninth high court (Japan has high courts for eight major jurisdictions). Others from a more bureaucratic perspective believed that litigation resulting from IP disputes could be accommodated by reinforcing the existing special division at the Tokyo High Court. A compromise solution was finally agreed on under the tentatively named 'Intellectual Property High Court' that will operate within the Tokyo High Court but with a high degree of independence. This new high court will deal with appeals of district court rulings over rights infringement lawsuits and lawsuits filed by those who are dissatisfied with the Patent Office's decisions on their applications for rights.

The decision was welcomed by the JPAA who had sided with an independent high court for intellectual property infringement lawsuits. The JPAA views the independence of personnel management and the mobilization of judges and examiners who are experts on matters dealing with intellectual property as a priority. The Intellectual Property High Court was seen as even more necessary because already in 2001 of some 440,000 applications for patent rights registration filed with the Patent Office, at least 20,000 complaints were filed by applicants against its decisions of refusal.

The Intellectual Property High Court emerged in a smooth transition because the numbers of divisions and judges had already increased slowly to accommodate the rapid accumulation of intellectual property related cases.

The IP High Court has 18 judges in four divisions, which is equivalent to a medium sized high court. The Judges are supported by 11 research officials. Ten of these have experience in the fields of machinery, chemistry or electricity as an examiner or appeal examiner of the Japan Patent Office (JPO) and one has experience as a patent attorney. These research officials are required as court officials to review the JPO Board decisions and to submit a fair and accurate report to the Judges.

Finally, to ensure maximum fairness, each of the four divisions of the IP High Court has a special division, the Grand Panel Division. The IP High Court is the court of second instance and the Supreme Court is the court of last instance in terms of legal interpretation, but the business community has requested the judiciary to form reliable rules and standards at an appeal stage so that they do not have to wait for decisions by the Supreme Court. For these reasons, the Grand Panel system was introduced – this happened in April 2004 with the amendment of the Code of Civil Procedure in order to provide unified opinions at the second instance level.

An Intellectual Property Nation

Former Prime Minister Koizumi made his Administrative Policy Speech on 4 February 2002, stating that:

> Japan has already produced world class intellectual property. I have set the national goal of strengthening the international competitiveness of Japanese industry through strategically protecting and utilizing achievements in research and creative activities as our intellectual property. To this end, Japan will establish the Intellectual Property Strategy Committee and forcefully promote necessary policies.[4]

The Intellectual Property Strategy Committee which was established in the Prime Minister's Cabinet Office published the Intellectual Property Policy Outline in July 2002. The Intellectual Property Headquarters (Director-General, Prime Minister Koizumi) was established in March 2003 and the Japanese government announced the Intellectual Property Strategic Program in July 2003. The key points are as follows:

1) The creation of IP:
    a) Expansion of national subsidies for the acquisition and management of intellectual property rights at universities.
    b) Abolition or amendment of the Employee's Invention Rules.
2) The protection of IP:
    a) Enactment of the Faster Patent Examination Law.
    b) Expansion of the patent subjects to medical practice.
    c) Study on the establishment of the Intellectual Property High Court.
    d) Study on the establishment of the International Intellectual Property Trade Committee (provisional name) to counter infringements.
3) The utilization of IP:
    a) Utilization of the trust system for management and mobility of intellectual property.
4) Expansion of the contents industry:
    a) Diversification of fund procurement methods and financial support.

5) Education:
   a) Education of attorneys specializing in intellectual property.
   b) Promotion of the establishment of a graduate school, Management
      of Technology (MOT).

Further Measures by the Industrial Property Administration to Modernize IPR: The Role of the Japan Patent Office (JPO)

Further measures by the Industrial Property Administration to modernize IPR include defining the role and responsibilities of the Japan Patent Office (JPO). The following measures are of particular significance in this regard:

● *Fast and accurate examination and judgement*

The realization of fast and accurate examination and judgement is a priority mission in industrial property administration as pointed out in the Intellectual Property Policy Outline. At patent examinations, in particular, the Japan Patent Office (JPO) will wrestle with central issues of ensuring promptness in granting rights while maintaining the high quality of examination.

● *Prompt granting of rights*

Among the various procedures related to industrial property, the JPO initially speeded up the filing of documents for formal examination and registration of industrial property.

   To speed up the patent examination, it is important to establish a broad system capable of meeting diversified needs since the necessity for examination speed differs for each applicant in each field of technology. The JPO has been in the process of organizing an efficient examination processing system because it is taken for granted that the examination starts as quickly as possible after an applicant submits a request for one. The JPO has also standardized an accelerated examination that significantly shortens the waiting time for applications by universities, small- and medium-sized enterprises and venture companies, patent applications that are also filed overseas, and patent applications that are certain to be executed. Promptness of granting a patent is particularly required in those cases, and the JPO has been actively promoting utilization of this accelerated examination to the applicants. The JPO has also set a quantitative target for the period from the application for an accelerated examination to notification of the result of the primary examination and has been processing the applications quickly. The JPO has been furthermore proactively executing circuit examinations with the objective of supporting small- and medium-sized enterprises, venture companies, and so on, in all areas of Japan.

   As for the design examination, the JPO has set a quantitative target for the period of the accelerated examination for those applications that particularly require early processing because the need for faster granting of rights has been higher as the product cycle becomes shorter. As for the trademark examination, the JPO supported the provision to complete the primary examination within 18 months for all cases. For applications for accelerated examination that urgently require granting of rights, the JPO has set a quantitative target for the period from the filing of the application to notification of the result of the primary examination and shall process them quickly (Tani, 2004).

Accurate Granting of Rights

The JPO is continuing to attempt to execute the procedures of the formality examination and industrial property registration properly and accurately, to achieve stable operation of the online reception system and take every precaution to prevent causing accidental inconvenience to the applicants. With regards to substantive examinations, the JPO is trying to achieve equal and stable granting of rights by reviewing the correction limits and other examination standards and by ensuring that everyone is thoroughly familiar with the examination standards. Specifically, the JPO shall thoroughly investigate prior art examples, issue proper judgement to ensure objectivity of the results of substantive examinations and reduce the number of reversals at subsequent trials for invalidation in order to be able to grant highly reliable and stable rights. The JPO has also been reorganizing the examination-related information system to establish an environment for prompt and accurate examinations.

Prompt and Efficient Intellectual Property Dispute Processing

The JPO has reviewed the relationship between the opposition system and trial for the invalidation system based on the Intellectual Property Policy Plan and submitted the bill to the ordinary Diet session in 2005 for unification into a new trial system for invalidation that has the benefits of both systems. The JPO is striving for a proper operation of the judgement system by simplifying and streamlining it and is reorganizing the functions required by the Intellectual Property Policy Plan in observance of the results of the subject bill at the Diet. The JPO is ensuring that anyone demanding a trial appeal has a thorough knowledge of the new system.

Support for Creation of Excellent Designs and Brands, and their Strategic Utilization

The significant elements in strengthening industrial competitiveness are creation, protection and utilization of attractive designs and brands as well as promotion of technological development. The JPO, therefore, has been studying the policy to promote utilization of design- and brand-related information in its possession and prepare a bill within this fiscal year in order to promote structuring and creation of attractive designs and brands. The JPO also studied the design and trademark systems to decide on specific policies for establishing an environment to offer products and services with higher value by utilizing attractive designs and brands that came into effect in 2005. The JPO reviewed the ideal method for protection of designs (which also involved a process of valuation) to be utilized in the network. The information-oriented society is rapidly progressing.

Measures against Unauthorized Copies in Japan

The JPO has been promoting stronger cooperation with the customs houses to effectively prevent inflow of unauthorized copies to Japan. This policy is already taking effect against Chinese companies that have been trying to export Japanese goods that infringe patents (Tani, 2004).

Proper Protection of Intellectual Property Overseas

Damages incurred from infringement of intellectual property rights by unauthorized copies and so on, are increasing mainly in the Asian regions and seriously impacting the activities of Japanese enterprises. The JPO will, therefore, firmly request preparation of laws, establishment of a system and improvement of its operations against unauthorized copies to the countries and regions in Asia through bilateral and multilateral negotiations and strengthen international cooperation concerning protection of intellectual property jointly with European countries and the United States (Taplin, 2005).

In 2005, the JPO proposed that because of an agreement made during a series of meetings held between Chinese intellectual property rights officials and the Japanese delegation led by the former Chairman of Honda Motor Company, Mr Yoshihide Munekuni, Japan will assist in instituting a number of anti-counterfeiting measures. These include sending Japanese experts to China to hold seminars showing Chinese enforcement agents how to distinguish counterfeit products from the genuine articles and providing past examples of illegally copied goods. Mr Munekuni told a press conference that he believed China was making progress in inhibiting the production of counterfeit goods because government ministries are working closer together to improve enforcement.[5]

International Accord and the IP System

An International accord and strengthening of efforts for cooperation in examinations makes it necessary to file the same application in overseas countries in order to acquire international rights, and applicants are suffering from the heavy burden of the extra procedure and expense. The responsibilities of patent offices in all countries are rapidly increasing because the same applications must be examined. The JPO shall promote cooperation among worldwide patent offices including correlative utilization of prior art search results and examination results with due consideration to the sovereignty of each country in order to make the application process smoother and to reduce the number of responsibilities. The JPO will lead discussions on the reformation of the Patent Cooperation Treaty (PCT) in the World Intellectual Property Organisation (WIPO) to enhance the efficiency of the system and the convenience for applicants, and take active measures in discussions on the Substantive Patent Law Treaty (SPLT) at WIPO to achieve international accord in the patent system. As for the designs, the JPO will endeavour to establish a system in which the 'primary examination is conducted within 12 months for all cases as provided in the Hague Agreement Geneva Act (adopted in July 1999) and review compliance to the Act. With regard to trademarks, the trilateral trademark meetings of the Japan Patent Office, the United States Patent and Trademark Office (USPTO) and Office for Harmonization in the Internal Market (OHIM) shall review issues regarding the indications of product and usage names, Vienna figure classification, and so forth. The purpose here is to reduce the burden of examination responsibilities of the three offices and to enhance convenience for the applicants, and to promote international accord in the trademark system.

The number of participating countries in the Protocol Relating to the Madrid Agreement Concerning the International Registration of Marks, an international system for trademark registration, has exceeded that of the Madrid Agreement and further increase is expected. It is inevitable that applications for trademark registration in the framework of this Protocol will increase and the JPO therefore will take active measures to achieve international accord in the trademark system to make the Protocol even easier for applicants to utilize. The JPO will also make efforts in conducting a review for the amendment of the Trademark Law Treaty (TLT) and substantive accord of the trademark system at the standing committee for laws on trademarks, designs and geographical displays at WIPO and strive to achieve international accord.

The Strategic Programme seeks to make universities more IP focused, speed up patent application examinations, increase anti-counterfeiting measures, support SMEs by reducing patent fees and subsidizing applications for foreign patents, and increase the number of IP professionals in Japan, namely lawyers (*Bengoshi*) and patent attorneys (*Benrishi*).

IP as a Policy Priority in Japan

From the above discussion it becomes clear that intellectual property has become a priority for Japanese policy makers at the highest level. Furthermore, Japanese industry, which is keen to protect itself from products produced by Asian competitors at lower cost, has shown signs of attaching more importance to intellectual property rights and their enforcement. In light of this, both government and industry have been keen to have a specialist patents enforcement system. As stated, patent law is very specialist and patent proceedings involve technical factual aspects. There is therefore a real advantage in having specialist courts to consider patent issues. With the IP strategy being of such importance at government level, this has now been achieved.

Recent Changes to the Japanese Patent Court System

The recent changes that have been made to Japan's patent procedure relate to the patent court system to make it more specialist; and to the way in which invalidity can be raised in proceedings and a patent revoked.

Past Procedures

While a patent may be granted by a patent office that is not to say it is certainly valid. It may not be new or inventive in light of prior art (or an argument) that has escaped the examiner. The claims may be too broad for the invention that has been disclosed, or the patent specification may otherwise not enable the invention to be performed. The invention may also not be industrially applicable. If a patent is found not to be valid, it cannot be infringed.

Previously, the validity of a Japanese patent could only be challenged in the Japan Patent Office (JPO). Indeed, the Court was not able to consider validity at all during infringement proceedings. If infringement proceedings were commenced, they would sometimes be stayed pending determination of the validity of the patent by the JPO. There were two procedures that could be carried out in the JPO – the opposition procedure and the invalidation appeal. On occasions, both procedures could be invoked against the patent, although the rules as to who was eligible to commence such proceedings and the time limits differed. Under the opposition procedure, it was open to anybody to oppose the validity of the patent within six months of the advertisement of the granted patent in the Patent Gazette. Under the invalidation procedure only an interested party was able to challenge the validity of the patent. There was no time limit for doing so under this procedure. Both procedures were conducted before a body of three or five examiners of the JPO. Under the opposition procedure a dissatisfied patentee could appeal the body's decision to the Tokyo High Court; the opponent, however, had no means to appeal. Under the invalidation procedure, either party could appeal the JPO's decision to the Tokyo High Court.

In 2000, the Supreme Court of Japan decided in *Fujitsu* v. *Texas Instruments* that validity grounds could be raised in infringement proceedings. If there were convincing grounds for considering the patent to be invalid, then the infringement action would fail. This was important as it allowed 'squeezes' to be raised, in which the patentee cannot argue on the one hand that the claim is broad enough to cover the alleged infringing article, if that breadth also means that it renders the patent invalid (for example by covering prior art). However, the Court still did not have power formally to decide that the patent was invalid – merely to dismiss the infringement action on the basis that the patent appeared to be invalid.

Infringement

The first instance court for patent proceedings in Japan is the District Court. Appeals from the District Court can be made to the High Court. A final appeal, on questions of law only, rests with the Supreme Court of Japan.

Previously, each District Court (of the 50) no matter how little expertise in IP matters it had, had exclusive jurisdiction to hear patent disputes in its district. However, from 1998, Tokyo District Court was given concurrent jurisdiction with eastern District Courts for all eastern areas of Japan and Osaka District Court was given concurrent jurisdiction with the western District Courts for all western areas. All patent disputes could still be heard in the other District Courts, however the majority of all patent cases were from that time heard in Tokyo or Osaka. These two District Courts have specialized divisions dealing with IP (for example, three in Tokyo) and the first tier courts therefore, in practice, became more specialized.

Appeals from any particular District Court lay to the regional High Court having jurisdiction over the district. This led to appellate courts with little expertise having jurisdiction over patent cases, however; again, after 1998 most appeals have been heard in Tokyo and in Osaka.

Recent Developments

As stated above, with the importance of intellectual property being acknowledged by both government and industry, there has been pressure for the Courts to become more specialized in dealing with, in particular, patent disputes. After much debate and discussion, consensus as to the way forward has been reached.

From April 2004 some major changes were made to the Civil Procedure law. First, all litigation relating to patents, utility models, circuit design rights and copyright in computer programmes is to be assigned exclusively either to the Tokyo District Court or Osaka District Court. Accordingly, it will no longer be possible to bring proceedings in other District Courts, which previously had concurrent jurisdiction with those two courts. The number of specialist divisions in Tokyo District Court has been increased to four to meet the increased demand. In addition, all appeals will now be heard by the Tokyo High Court, which has by far the most specialization and expertise in the IP field. The number of judges in the IP division of the Tokyo High Court was increased from 16 to 18 to meet the additional demands.

Secondly, in the Tokyo High Court, a grand panel system was introduced to ensure consistency of High Court decisions. When cases raising the same issues are pending in the High Court, they will be heard by a panel consisting of five leading judges with IP expertise.

Thirdly, 140 technical advisers have been appointed to assist the High Court and District Courts on technical matters. The technical advisers are university professors or researchers in public or private organizations who will be able to assist the court with their expertise on a part-time basis when required. This is in addition to the full-time research officials who are already employed by the courts as mentioned above.

Since April 2004, therefore, substantial practical changes have been made to the patents court system to make it more specialist and hence reliable. However, in 2005 an amendment to the law came into effect creating a new IP High Court, as part of the Tokyo High Court. This, in fact, was something of a compromise, as some had wanted an entirely independent IP High Court as mentioned above. Nevertheless, the amendment to the law and the formal creation of the court has had the effect of enshrining in law the new position with respect to IP specialization, the formality being viewed as appropriate in light of the importance being attached to IP rights at present in Japan.

Another major amendment made to Japanese law that came into effect in 2005 was to allow invalidity to be raised formally as a defence to infringement. Rather than arguing merely that the patent appears to be invalid and thus should not be enforced, which has been the position since *Fujitsu* v *Texas Instruments*, the Court has been able formally to decide that the patent is invalid. This decision however will be binding only on the parties, as the power to revoke the patent will remain with the JPO.

In addition, the former position whereby there were two ways of challenging validity in the JPO has been abolished. There is now to be only one procedure, a revised invalidation procedure. Under this procedure, any person can challenge the validity of the patent, and there are no time limits for doing so. The decision of the JPO may be appealed to the IP High Court.[6]

Conclusions and Implications

In conclusion, it can be stated that these important changes have made the Japanese patent system more specialized and better able to cope with this area of law with its complexities. In terms of implications for IP development in Japan, the changes to the procedures for attacking the validity of patents are also important as they render proceedings fairer and more efficient. No doubt, there will be many more developments in the future, but these changes will help to ensure that Japan has a fast and reliable patent enforcement system. These changes are further supported by the amendment made to Section 4 of Article 35, which finally provides a balanced law that serves to provide proper remuneration for the architects of invention, the inventors themselves, while balancing the contribution of the employers. For, without inventive and creative people there is no innovation in any society; over time, both economy and society perish without creative talented people.

Notes

[1] See Taplin (2003b). Much of this material has been gathered from interviews with the contributors to my books on the subject or from the Japanese judiciary (see below and last footnote). Academic literature is lacking in the English language particularly concerning Japanese IPR because of language and cultural difficulties.

[2] Hiroshi Okuda, Chairman of the Japanese Business Federation, speaking in 'Settlement in LED lawsuit could spark surge in lawsuits by inventors'. January 2005. Available at http://www.asahi.com.

[3] The draft laws were passed in April 2003, see the Appendix in Taplin (2003a).

[4] This was taken from the Japanese Administrative Policy Speech made by Prime Minister Koizumi on 4 February 2002 (in Japanese).

[5] April 2005. This article was cited in a Jiji Press (Japanese) news bulletin on 16 June 2005.

[6] The author would like to thank Judge Ryuchi Shitara, Presiding Judge of the Intellectual division of the Tokyo District Court, for his assistance with material concerning the IP High Court and employee rights to compensation.

References

Japan Times (2004) 24 February.

Nishizawa, A. (2006) *Innovation and Business Partnering in Japan, Europe and the United States* (London: Routledge).

Tamai, K. (2003) Employee's Rights to Invention – A Changing Situation, in: R. Taplin (Ed.) *Exploiting Patent Rights and a New Climate for Invention*, pp. 19–26 (London: Intellectual Property Institute).

Tani, A. (2004) Redefining brand valuation within the Japanese context, in: R. Taplin (Ed.) *Valuing Intellectual Property in Japan, Britain and the United States*, pp. 110–130 (London: RoutledgeCurzon).

Taplin, R. (Ed.) (2003a) *Exploiting Patent Rights and a New Climate for Innovation in Japan* (London: Intellectual Property Institute).

Taplin, R. (2003b) Overview: Japanese attitudes to litigation and IPR, in: R. Taplin (Ed.) *Exploiting Patent Rights and a New Climate for Innovation in Japan*, pp. 1–9 (London: Intellectual Property Institute).

Taplin, R. (2004) Introduction, in: R. Taplin (Ed.) *Valuing Intellectual Property in Japan, Britain and the United States* (London: Routledge Curzon).

Taplin, R. (2005) In defence of IP; Japan's new approach to patent protection, *Patent World,* Issue 171, pp. 20–22.

Trenton, A. (2005) Developments in patent enforcement procedure in Japan and England, in: R. Taplin (Ed.) *Risk Management and Innovation in Japan, Britain and the United States*, pp. 77–90 (London: Routledge).

Young, T. (2004) Technology transfer from US universities: the need to value IP at the point of commercialization, in: R. Taplin (Ed.) *Valuing Intellectual Property in Japan Britain and the United States*, pp. 20–33 (London: Routledge Curzon).

From Vertical to Horizontal Inter-Firm Cooperation: Dynamic Innovation in Japan's Semiconductor Industry

YOSHITAKA OKADA

Introduction

In the 1980s, Japanese semiconductor manufacturers came to dominate the world DRAM (dynamic random access memory) market with the 64-kilobit (Kb) in 1982, the 256 Kb in 1984, and the 1-megabit (Mb) in 1988. Japan's world market share for all types of semiconductors first approached that of the USA in 1985, passed it in 1987 (Japan at 48 per cent vs. the USA at 39 per cent), and peaked in 1988 with Japan at approximately 51 per cent vs. the USA at around 37 per cent (Okada, 2000). Japan's share then gradually declined to 29 per cent in 1998 while the USA recorded 52 per cent (Handotai Sangyo Kenkyusho, 2000). Although some scholars argued that the decline was due in large part to the burst of the bubble economy in 1992 (Chon, 1997), Japan's world market share had already begun to decline long before then.

Japanese success in the semiconductor industry owed a great deal to the development of the very large-scale integrated circuit (VLSI) in 1977, made possible by the success of the government-subsidized VLSI Cooperative in generating inter-firm cooperation[1] among competing large-sized firms. Another important factor explaining this particular industry's success was the dynamic vertical inter-firm cooperation for research and development (R&D) between equipment and semiconductor manufacturers (Okada, 2000). Such cooperation became possible only when firms interacted with a shared understanding of needs for continuous improvements, of benefits for cooperation, and of a relationship based on mutual trust. This relationship promoted cooperative learning – a process of mutually increasing the exchange of rich technological information among partners, promoting learning and advancing technological capability. This type of relationship generates synergetic effects (Okada, 1999), facilitates goal attainment, and brings positive long-term gains among partners. It occurs both within and among firms, governments and technology supporting organizations (TSOs), such as universities and research institutes. Meanwhile, the firms involved compete fiercely for market share.

However, and despite these apparently mutually beneficial patterns of strategic interaction, Japanese semiconductor manufacturers lost international competitiveness during the 1990s, failing to maintain their technological leadership. This study explains how this situation came about and how Japanese semiconductor manufacturers attempted to develop an innovation-based response.

Research Questions and Propositions

This study addresses the following research questions:

- Why did the Japanese semiconductor industry lose international competitiveness, despite the existence of what appeared to be such an effective national system of innovation?
- What came to replace the system after the so-called 'lost decade' of the 1990s?
- How different was this new system from the old one?
- How can we explain the transformation?

In addition, this study advances the following propositions:

Proposition 1. The old techno-governance structure characterized by vertical intra- and inter-firm cooperation, though fitting well to Japanese institutions, entrapped Japanese companies in a DRAM-based technological trajectory that restricted their choices for decision making and made them incapable of coping with path-disturbing contingencies.

Proposition 2. The withdrawal of both macro- and micro-level strategies led to the disintegration of the old techno-governance structure in Japan.

Proposition 3.   In order to develop a new type of large-scale integrated circuit (LSI) which contains a system (called the system LSI), companies adopted micro-level strategies designed to promote horizontal intra- and inter-firm relations. The latter resulted in a complex mix of cooperation and competition in modularized product areas. In terms of macro-level strategies, they revived the traditional sense of cooperation, some with government subsidies and some without. The characteristics of the new techno-governance remained within the institutional orientation of mixing cooperation and competition.

In addressing these research questions and testing these propositions, this study first develops the concept of techno-governance for innovation by integrating the concepts of cooperative learning and North's institutions (North, 1989, 1990) together with Lundvall's discussion of national systems of innovation or NIS (Lundvall, 1988, 1992). It then explains the decline of the Japanese semiconductor industry as evinced in the disintegration of the old national system and the rise of a new one characterized by a new type of strategic mix between cooperation and competition with newly shared interpretation of what 'cooperation' might mean. In developing this analysis, the contribution illustrates how these new structures for innovation built on established institutions and a strategic emphasis mixing cooperation and competition.

Context: Techno-Governance Structures for Innovation in Japan

As a first step in contextualizing this discussion, the concept of 'techno-governance structure' is explored in relation to innovation in semiconductor manufacture in Japan.

Cooperative Learning and Technological Innovation

Technological innovation is one of the leading factors in maintaining industrial development and competitiveness. It requires knowledge and information accumulation within as well as mobilizations among diverse participating actors. According to the national system of innovation (NIS) argument, companies and technology-related actors engage in interactive learning, searching and exploiting through coordinated routines and feedback channels with each other (Lundvall, 1988, 1992; Lundvall & Maskell, 2000). Among actors, companies are considered to be the most effective accumulators and users of knowledge and information. Technology-related actors such as governments and technology supporting organizations or TSOs are also regarded as playing vital roles in promoting interactive learning and supporting and stimulating technological innovation. Needless to say, cooperative learning with shared values and commitment to long-term relations – as was the established pattern in Japan – can generate much more synergetic effects than simple interactive learning (Okada, 1999, 2001). Since technology is the outcome of learning differentiated by the ways actors interact and accumulate information, it is understood not simply as mechanical or functional information but rather as the embodiment of knowledge in people, influenced by diverse socio-economic factors. Hence, an important path of inquiry explores how companies and other technology-related actors interact with each other in response to shifting socio-economic factors (Okada, 2006a).

Intra- and Inter-Firm Interaction between Technology-related Actors

Williamson (1975, 1981, 1985) argues that hierarchy and markets are two extreme forms of business transactions. Hierarchy, on the one hand, internalizes all transactions within an organization and enhances innovative capability through effective organizational control and coordination. At the other extreme, a market consists of arm's-length transactions and short-term contracts, enabling decision makers to introduce flexibility and openness for quicker changes, though it fails to facilitate effective interactive learning without shared tacit knowledge (Metcalf, 1998). Owing to his theoretical focus on contracts, however, Williamson fails to focus on intermediate forms between hierarchy and markets, whose relations – as found in Japan – are based on long-term cooperation. Such intermediate forms, mixing the characteristics of both hierarchy and market, generate a distinctive type of relations contributing to technological dynamics (Powell, 1990). Hence, and for some functions and contingencies, a company may rely on internalized operations while for others it may choose long-term or arm's length relations with other companies. The manner in which a company structures its intra- and inter-firm relations becomes one important factor for increasing its innovative capability and competitiveness in the market in addition to informing its attempts to minimize transaction costs.

Besides, a company's relations with technology-related actors also require considered attention, given that interactions with diverse actors who bring different ideas and backgrounds to the relationship serve to disseminate information, promote learning and stimulate innovation. In bringing different sets of information, networks, capabilities and funding sources to the interaction, governments and TSOs are often cited as important actors in differentiating the innovative capabilities of companies, industries, and nations – and even in creating distinctive varieties of capitalism (Nelson & Winter, 1977, 1982: Nelson, 1993; Hall & Soskice, 2001; Streeck & Yamamura, 2001). Scholars of the institutional arrangements perspective even argue that diverse types of coordinating and control mechanisms compensate for the failure of the market in exchanging tacit knowledge and the unmet needs of intra- and inter-firm relations (Hage & Hollingsworth, 2000). Hence, technology-related actors play important roles in substituting vital missing resources and linkages, contributing especially to the harmonization of diverse actors.

When harmonization among companies and technology-related actors comes to show a distinctive and repeated pattern, it can be said that they form a techno-governance structure: that is, a system networked and harmonized among companies, governments and TSOs, designed for allocating limited resources, disseminating technological information, promoting interactive learning, stimulating innovation, and influencing the transformation of industry. Japan especially has a long and distinctive history of emphasizing cooperation for industrial development: for, the ways in which diverse actors are harmonized strongly influences the capability of any one country to generate a competitive advantage in technology (Okada, 1999, 2001, 2006a).

Institutional and Non-institutional Contingencies in Techno-governance Structures

The effectiveness of the techno-governance structure is (as Williamson argues) a matter of the 'fit' between the structure and its contingencies (Williamson, 1985). The contingencies at issue here comprise two types: institutional and non-institutional.

North (1989, 1990) defines 'institutions' as the formal and informal rules of games and the enforcement characteristics of rules in human interaction. They develop the cognitive orientation of individuals and organizations, path-dependently restrict their options and choices and propagate repeated patterns of interaction. Institutions are different from actors who have the capacity to change as well as to maintain a set pattern of practices characterizing institutions. Actors also stimulate new and innovative ideas, induce sub-optimal solutions within the acceptable range of restrictions, and contribute to gradual changes (Metcalf, 1998). This means that even path-dependent and restrictive institutions have their creative aspects, enabling actors to develop diverse sets of mechanisms and to coordinate them in developing technological dynamics. One of the key limitations of the national system of innovation or NIS argument is its failure to separate institutions from technology-related actors, resulting in blurring the influence of long-lasting values and norms as well as the nature of interactions between diverse actors. Foci on the complex interactions between institutions and actors are vital; not least in understanding changes in Japan's techno-governance structure during the 1990s.

Among many non-institutional factors, product market conditions and historical incidents are most significant in influencing company behaviour. This is because technological innovation without infiltration into the market cannot be properly considered as innovation per se (West, 1992). This is also because certain historical incidents may generate a self-reinforcing or antagonistic – and often irreversible – momentum, compelling other companies and even competitors to follow blindly. These two types of institutional and non-institutional contingencies are opposite in their characteristics. Institutions are persistent and evolve very gradually, while product market conditions and historical incidents usually occur sporadically and inconsistently.

Hence, and as Figure 1 shows, understanding the complex process of technological innovation requires us to focus on two sets of relations: one formulating the techno-governance structure, and the other as contingencies to the structure. The former consists of: 1) intra- and inter-firm relations; 2) companies' relations with technology-related actors that directly influence technological innovation; and 3) companies' relations to diverse types of coordination mechanisms. The latter consists of companies' relations to: a) institutions that persistently restrict options

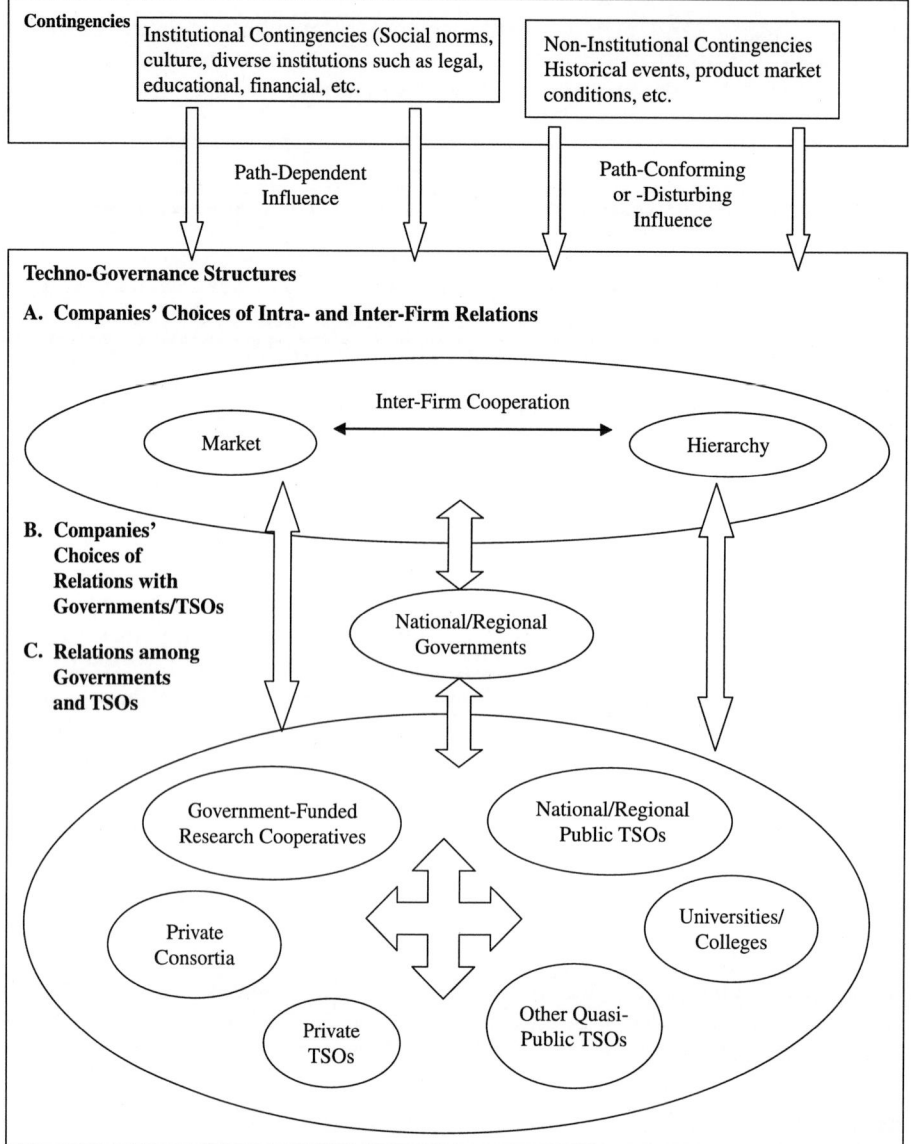

Figure 1. Techno-governance structures and contingencies

and choices for decision makers; and b) product market conditions and historical incidents. The effectiveness of the techno-governance structure identified by the former clearly differs in terms of its fit to the contingencies identified by the latter.

Contingencies, Strategies and Changes in Techno-governance Structures

Historical incidents and product market conditions can be interpreted as being either favourable or unfavourable for a techno-governance structure. When they are favourable or less disturbing, companies accommodate them as path-conforming contingencies (see Figure 1) with slight alternations within a framework of well-functioning techno-governance and path-dependent institutions. Conversely, when they are unfavourable or extremely turbulent (that is, path-disturbing) contingencies, companies may drastically change their intra- and inter-firm relations and their relations to technology-related actors by introducing new strategies for survival. These changes can conceivably result in altering the existing techno-governance structure and even institutions, though such alterations may remain limited owing to the restrictive nature of institutions.

In order to accommodate with changes in contingencies, companies develop micro- and macro-level strategies.[2] Micro-level strategies can be managed within a company's decision making and are designed to change relations inside the firm and with other firms, governments and TSOs. They are much easier to decide on and implement than macro-level strategies, since they are essentially a matter of decisions made by each company. However, and particularly because of the narrow range of activities that a given company can influence, self-initiated micro-level strategies can become less effective, and especially when non-institutional contingencies are turbulent.

In contrast, macro-level strategies employ collective actions, which in many cases involve governments and various TSOs directly and indirectly related to technology innovation. They influence government policies and the behaviour of public and quasi-public TSOs, often resulting in establishing coordinating and control mechanisms to orchestrate resources (Casper, 2000). Through collective macro-level strategies, companies may even try to transform institutions so that micro-level strategies become more effective and better supported by macro-level environments.

In a sense, a company's efforts to cope with turbulent contingencies clearly appear in micro- and macro-level strategies and, if successful, result in changes in the techno-governance structure. Enormous energy is required to release companies from initial self-reinforcing mechanisms. If leading and powerful companies adopt similar micro-level strategies and a significant number of others follow, those strategies may result in reformulating the techno-governance structure and reducing resistance against changes. However, this becomes extremely time-consuming and severe market competition may not give individual companies sufficient time to survive. Developing a solution to such difficult situations often involves macro-level strategies and collective actions. If they work effectively, they relatively quickly change the techno-governance structure, reduce the costs of obtaining information, promote technology transfer with trial and error; in addition, they provide a favourable environment for micro-level strategies (Okada, 2006a).

The question thus arises as to how and to what extent the old structure under path-disturbing contingencies becomes dismantled, de-institutionalized and replaced by a new emerging structure (DiMaggio, 1988). How well does the new structure fit with contingencies? Analysing changes in the micro- and macro-level strategies of semiconductor companies provides some answers to these questions and furthermore helps explain the nature and outcome of shifts in dynamic innovation in Japan's semiconductor industry.

Empirical Propositions

The old techno-governance structure that existed before the 1990s in the Japanese semiconductor industry was characterized by lifetime employment of core staff together with long-term vertical cooperative relations with equipment manufacturers suitable for DRAM developments and production. However, in the 1990s this established structure faced turbulent historical incidents and product market conditions – so turbulent that they caused the decline of the industry in Japan.

Some forces underlying this drastic decline were as follows:

1) The large storage capability of computer chips weakened demand for the DRAM.
2) Samsung of South Korea started dominating the high-end DRAM market, while Micron Technologies in the USA started occupying the low-end market.
3) Technological leadership shifted from the development of the DRAM by Japanese companies to that of complex function-specific VLSIs by US companies (Okada, 2006b: 39–42).

Consequently, the centres of semiconductor industrial dynamics and advancements in technology shifted away from Japan, suggesting that the old techno-governance structures could not be relied on as a source of international competitiveness.

Proposition 1. The old techno-governance structure characterized by vertical intra- and inter-firm cooperation, though fitting well to Japanese institutions, entrapped Japanese companies in a DRAM-based technological trajectory that restricted their choices for decision making and made them incapable of coping with path-disturbing contingencies.

Macro-level collective actions aided by governments and TSOs – which in the past had been sources of technological dynamics – came to be discontinued, owing to a combination of political and economic factors in the 1980s. Along with the withdrawal of old micro-level strategies in the 1990s, the discontinuity also of old macro-level strategies caused the disintegration of the old techno-governance structure.

Proposition 2. The withdrawal of both macro- and micro-level strategies led to the disintegration of the old techno-governance structure in Japan.

In their attempts to survive, Japanese companies engaged in a micro-level strategy of developing technologies for the system LSI and to this end changed to intra- and inter-firm relations based more on horizontal cooperation suitable for the development. Delayed action made it necessary for Japanese companies to modularize their organizational structures, resulting in a complex mix of inter-firm cooperation and competition based on modularized product areas. In terms of macro-level strategies, Japanese companies also developed research cooperatives, some with government subsidies and some without, reviving the traditional form of cooperation. A new structure of techno-governance came to be developed based on horizontal cooperation with newly shared interpretation of 'cooperation'. However, context was still given by the traditional institutional orientation of mixing cooperation and competition.

Proposition 3. In order to develop a new type of large-scale integrated circuit (LSI) which contains a system (called the LSI system), companies adopted micro-level strategies designed to promote horizontal intra- and inter-firm relations. The latter resulted in a complex mix of cooperation and competition in modularized product areas. In terms of macro-level strategies, they revived the traditional sense of cooperation, some with government subsidies and some without. The characteristics of the new techno-governance remained within the institutional orientation of mixing cooperation and competition.

Drawing on recently conducted research, the above propositions will be examined in the following sections.

The Limitations of Old Structures based on Vertical Cooperation

Japan's past successes had encouraged its engineers to believe firmly in the continuity of established practices in their attempts to advance the forefront process technology for the DRAM and make extra efforts to increase yield. Such technological lock-in blinded the semiconductor manufacturers against recognizing alternative paths to develop different devices such as the system LSI. Of particular significance is a company environment characterized by the strong sense of cooperation within each section that generated factionalism, separating production from design and design from algorithms and architecture. Managers at one venture firm (MegaChips) and the Joint Research Center for Atom Technology (JRCAT) even expressed the belief that a concept such as the system LSI – which requires the amalgamation of many areas of specialization – would never have emerged from this type of environment.

The same criticism can be applied to vertical inter-firm relations. In order to develop advanced equipment and obtain a higher yield out of silicon wafers, semiconductor companies nurtured the highly advanced craftsman-type technological capabilities of equipment, parts, and material producers. Semiconductor manufacturers became both centres of new information and innovative stimulation and providers of expensive facilities for testing. Significantly, they guided equipment manufacturers to develop highly relation-specific technologies, giving long-term partners priority in the development of any new equipment and ended up excluding new competitors from business opportunities (Okada, 2006b).[3]

Thus, intensive vertical intra-firm sectionalism and inter-firm cooperation – though fitting well to the Japanese institutional emphasis on cooperation and commitment to long-term relations – sustained an exclusive system of cooperative learning and encouraged a form of lock-in to a narrow technological perspective. The pursuit of such micro-level strategies in the spirit of the old techno-governance structure made Japanese semiconductor companies unaware of changes taking place outside of this structure and limited the range of their technological development.

The Disintegration of the Old Structure: Withdrawal of Macro- and Micro-level Strategies

Until the 1980s in Japan, TSOs and governments had functioned as technological information disseminators and innovation stimulators to companies. They uplifted the technological capability of companies, intensified competition with more value-added products, and stimulated industrial transformation. As a matter of fact, the development of the Japanese semiconductor industry was initiated by experiments conducted in government research institutes: the Telecommunication Research Institute (TRI) in 1951 and the Electro-Technical Laboratory in 1960. After becoming a research institute of a public corporation in 1952, TRI initiated a sequence of joint projects working on developing very large-scale integrated circuit (VLSI) and from 1975 initiated the development of the VLSI Cooperative (1976–79). In cooperation with semiconductor manufacturers, TRI succeeded in developing the world's first DRAM of 64 Kb (1977), of 128 Kb (1978), 256 Kb (1980), 1 Mb (1984), and 16 Mb – according to the last announcement made in 1987 (Okada, 2000). The privatization of the public corporation, however, forced the leading contributor to memory development to concentrate on telecommunications-related research, withdrawing from the field of semiconductors and move out of the old techno-governance structure.

The US-Japanese semiconductor disputes, begun in 1977 and aggravated in 1985, also had a devastating influence on the disintegration of the old governance structure in Japan. A series of incidents in the disputes forced the Japanese government to a series of drastic measures. These included:

1) shifting their projects to basic research, creating a wide gap between its research activities and the needs of companies (Fong, 1998);
2) forbidding any government research related to the silicon-based semiconductor; and[4]
3) turning down a project to nurture the system design capability of Japanese national universities in complex function-specific LSIs in order to compete against the USA.[5]

The concomitant disputes rendered the Japanese government incapable of soliciting cooperation among companies and TSOs on silicon-related research. In respect of 2) above, the situation became exacerbated as national research institutes even stopped funding internally conducted silicon-based semiconductor research. This was because of prohibitive research costs and because of a loss of research dynamics caused by a lack of researcher mobility to companies (Toray Keiei Kenkyusho, 1997). As a result, Japanese semiconductor companies started finding their relations with these TSOs and governments less useful to their research during the 1980s.

In the mid-1990s, a similar disintegration took place also at the micro-level. During this period, relations with equipment manufacturers came to be severed. The reason was rather simple. Joint equipment development requires intensive exchange of technology and information. However, when a partner company does not have the capability to purchase equipment, its manufacturer has to sell newly developed equipment to the partner company's competitors. During the 1970s and 1980s, ample orders and contractual agreements enabled Japanese semiconductor companies to delay the sale of newly developed equipment to competitors for one or two years (Okada, 2000).

However, during the 1990s equipment was sold to Taiwanese and Korean semiconductor companies almost immediately or at most six months after being developed. This meant that Japanese semiconductor manufacturers, through joint equipment development, were giving their technological expertise away to their international competitors. To illustrate some of the consequences of this decision: NEC sold its shares of Ando Electric stock to Yokogawa Electric and Hitachi turned its internal equipment production sections into independent companies and drastically reduced its internal or joint development.[6] In order to survive, a limited number of equipment manufacturers started to set up their own clean rooms and simplified mini-testing lines, and many even began to engage in close cooperation with Korean, Taiwanese and US companies instead of with Japanese companies. For example, after losing its close cooperative relationship with NEC, Ando Electric started cooperating with Samsung, which needed such an arrangement in order to reduce its dependence on Applied Materials.[7] In the late 1990s, for the first time even Japanese *keiretsu* companies had to face the severance of cooperative relations. Semiconductor manufacturers were forced to withdraw their old micro-level inter-firm strategies.

To summarize this section of our discussion: compounded by the companies' earlier withdrawal of their old macro-level strategies with the government and TSOs in the mid-1980s, the discontinuance of inter-firm cooperation with equipment manufacturers in the late 1990s decisively led to the disintegration of the old techno-governance structure relevant to Japanese semiconductor manufacturing (Okada, 2006b).

The Development of New Structures with Horizontal Cooperation

Against the background of the analysis presented so far in this study, it is possible to explain the shift to new structures of cooperation in the context of the Japanese semiconductor industry.

New Micro-level Strategies

Losing international competitiveness, Japanese semiconductor companies shifted their strategic foci from the DRAM to the system LSI, which is a chip that contains diverse system devices such as microprocessor, logic, memory, and so on. It comprises either one chip (system-on-a-chip or SoC) or one package of a chip shape (system-in-a-package or SiP). Although these companies had actually begun building up their capability of the SiP system LSI early on, it was only after their 1996 withdrawal from all-purpose DRAM production that they started focusing fully on the system LSI. For example, in 1996 Toshiba dismantled a vertically split R&D structure, focused on a specific device such as the DRAM or microprocessor, and gathered all related researchers for product development under one roof. NEC officially established a System LSI Divisional Headquarters in 1997. Mitsubishi's SiP system LSI strategies with embedded memory (eRAM – which mounts a DRAM on a logic LSI) began generating some hit products in 1998, and emerged in 2000 as the world leader in eRAM technology and packaging with a large number of pins.

Furthermore, some companies introduced an organizational structure for designing a system LSI within a limited time, concurrently mobilizing and horizontally coordinating about 1,000 people within and outside the company (Okada, 2006c: 116–117). Hence, by shifting their foci to the system LSI, Japanese semiconductor manufacturers came to develop intra- and inter-firm horizontal cooperation and coordination. However, and since difficulties to reduce production costs and increase a yield in developing SoC continued to trouble them, SiP systems remained the dominant products.[8] Besides, the biggest problem of identifying marketable-design-content in the system LSI continued to torment them.

However, unfavourable market conditions did not wait for Japanese semiconductor companies to solve their product, organizational, and technological problems. Companies were forced to lay off workers from 1998 onwards, discontinuing one of the core practices (lifetime employment) designed to maintain intra-firm harmony and cooperation (Okada, 2006c: 111). To increase their organizational flexibility, Sony (1994), Matsushita (1997), Hitachi (1998), and Toshiba (1999) introduced a new 'in-house company' system for separate units within a corporation (hereafter referred to as corporate headquarters), each with its own independent decision-making authority, personnel system, and cash flow accounting. The primary objectives of such restructuring were improved cost efficiency and management flexibility (Okada, 2006c: 118).

Meanwhile, several revisions of Japan's Commerce Law led to the propagation of another organizational system. In 1997, the Japanese government revised the law to facilitate mergers and acquisitions (*gappei seido*), stock exchanges and transfers (*kabushiki kokan/kabushiki iten seido*) in 1999, and (in 2001) the breaking up and spinning off of companies, known as the company separation system (*kaisha bunkatsu seido*) (Nagase, 2004: 146–156). Both the in-house company system and the company separation system created a basis for modularized company operations. Furthermore, it introduced more market-oriented relations among corporate headquarters, in-house companies and spun-off subsidiaries.[9] Overall, it generated organizational flexibility and fluidity. Consequently, Sony (1999, 2001), NEC (2000, 2002), Hitachi (2001, 2002), Mitsubishi (2001), and Matsushita (2002) divided their companies into several units, modularizing these units into in-house companies, some of which they later spun off as subsidiaries (*bunshaka*) (Yoshikawa, 2002: 2–42; Nagase, 2004: 133–136; Okada, 2006c: 118).

However, the relatively smaller size of the modularized and separated units also reduced each unit's financial and technological capabilities, rendering them too weak to compete internationally. This condition unexpectedly increased their mobility, creating opportunities for these units to be divested or merged with, acquired from, or aligned with other companies, and especially competitors. Following established practice, the preferred partners of Japanese companies were Japanese – partly in response to problems experienced with international alliances during the 1980s,[10] and partly because in technological terms Japanese companies had become less attractive to foreign competitors.

In 1999, NEC and Hitachi established Elpida Memory, a 50:50 joint venture designed to preserve production capability in the all-purpose DRAM area in Japan. Past successes in Japanese inter-firm cooperation usually involved a vertical relationship between a large-scale producer and a small-sized supplier, with unequal power relations and complementary specialization. This new alliance, however, involved two partners of equal power, both of whom boasted advanced technological capability as well as strong employee pride in and loyalty to their respective parent companies. Naturally, the alliance suffered from conflicts and difficulties in decision making regarding the location of factories within Hitachi or NEC, and about basic technologies and the system for sales (*Daiyamondo*, 2002).

Finally, in 2002 the new joint venture company decided to bring in a president from outside, hoping to resolve its internal conflicts. Despite the new framework of horizontal cooperation, old habits of vertical cooperation interfered with the decision making and daily operations of the new company. The institutional persistence of old and established concepts delayed decisive actions.

Sony, which placed less emphasis on traditional long-term relations, was one of the most active companies to switch partners quickly as alliance conditions changed. It aligned with Oki Electric and Fujitsu (in 1996 and 1998, respectively) in order to develop a DRAM embedded system LSI. In 1999 it again switched a partner to Toshiba in order to develop a central processing unit (CPU) for the PlayStation II, establishing a joint venture with 51 per cent owned by Toshiba (Press Journal, 1999). Then, in 2001, it announced alliances with Toshiba and IBM to develop the next-generation system LSI process technologies, and with Toshiba to develop the next-generation design technologies. In the liquid crystal display sector, it chose Samsung as a joint venture partner in 2003 (Okada, 2006c: 120). It can be argued that Sony enhanced its semiconductor capability by carefully choosing the right partners for each period. In the past, such frequent switching of partners had not been an accepted practice of Japanese inter-firm relations. However, amid the then semiconductor manufacturing crisis, unstable inter-firm relations of this type became a common phenomenon.

Mitsubishi switched its system LSI partner from Texas Instruments (TI) to Matsushita in 1998, with whom it engaged in a joint development (1999) and production (2000) of a new system LSI (Press Journal, 2000). Despite its success with Matsushita, Mitsubishi again switched its partner to Hitachi to merge their system LSI businesses and established Renesas Technology. Despite Fujitsu's loss in its battle against Toshiba to maintain its alliance with Sony, it successfully renewed relations with Toshiba to develop and produce a system LSI in 2002 (Press Journal, 2002). Meanwhile, in 2002 NEC made its semiconductor division an independent company, but chose not to align with any other company (Okada, 2006c: 120). In sum, and as shown in Table 1, the Japanese system LSI sector now

Table 1. New inter-firm alliances in the memory, system LSI and other areas

Product area	Company 1	Company 2	Product	Year	Comments
Memory and System LSI	NEC	None	system LSI	2002–	NEC Electronicks established
	NEC	Hitachi	DRAM	1999–	Elpida Memory established
	Mitsubishi	Hitachi	flash memory	1994–	
			1Gb DRAM	1997–1998	
			system LSI	2002–	Renesas Technology established
	Mitsubishi	Elpida Memory	DRAM	2002	DRAM section sold to Elpida Memory
	Mitsubishi	Matsushita	system LSI	1998–2000	
	SONY	Oki Electric	system LSI	1996–98	
	SONY	Fujitsu	system LSI	1998–99	
	SONY	Toshiba	system LSI for Play Station II	1999–	
			next-generation system LSI design technology	2001–	
			next-generation system LSI process technology	2001–	including IBM
	Fujitsu	Toshiba	DRAM	1998–2001	
			system LSI joint development	2002–	
	Matsushita	Toshiba	flash memory	1999–	including San Disk
Display	NEC	Mitsubishi	display business	2002–	NEC Mitsubishi Visual System established
	Hitachi	Fujitsu	plasma display	2003–	Fujitsu Hitachi Plasma Display established
	Hitachi	Toshiba Matsushita	IPS liquid crystal display	2004–	IPS Alpha Technologies established
	Matsushita	Toshiba	display business	2001–	Toshiba Matsushita Display Technology established
	SONY	Samsung	Liquid crystal display	2003–	S-LCD Corporation established

				Year	
Others	NEC	Matsushita	3rd generation mobile telephone	2001–	
	NEC	Toshiba	Satellite business	2001–	
			MRAM	2002–	
	Hitachi	Matsushita	home appliance digital network	2001–	
	Hitachi	Fuji Electric	power semiconductor	1999	Fuji Hitachi Power Semiconductor established
	Mitsubishi	Toshiba	3rd generation mobile telephone	2002–	
			transformer and electronics distribution business	2001–	
	SONY	Matsushita and others	blue ray disc for the next-generation DVD	2002–	Competition for a global standard
	Toshiba	NEC and others	high-definition DVD for the next-generation	2002–	Competition for a global standard

Source: Okada (2006c: 121).

has three key groups: unaligned NEC, Hitachi-Mitsubishi-Matsushita, and Toshiba-Fujitsu-Sony, although Matsushita's position in the alliance is not clear.

Such turbulence in the system LSI sector served to separate co-operators and competitors clearly from each other. The reality, however (and as shown in the sections of display and other product areas of Table 1), is far more complex than what might be depicted by reference to the memory and system LSI section. To illustrate, the following are some of the company alliances that emerged: Mitsubishi/Sharp on semiconductor standardization (1998); Toshiba/Matsushita to establish Toshiba Matsushita Display Technology (2001); Toshiba/Matsushita/Hitachi to establish IPS Alpha Technology (2004);[11] Toshiba/NEC on a magnetic random access memory (MRAM) (2001); and NEC/Matsushita on a third-generation mobile telephone (2001). Hitachi also exhibited a penchant for complex inter-firm relations. Some of its alliances are with Fujitsu and Sanyo on mobile telephone music delivery, competing against a Sony-Matsushita alliance; another is with NEC to establish Elpida Memory; another is with Mitsubishi to establish Renesas Technology; one is with IBM on hard disk drives; one is with Fujitsu to establish Fujitsu-Hitachi Plasma Display; another is with Fuji Electric and Meidensha on power semiconductors; and yet another is with Toshiba in the nuclear fuel business (*Daiyamondo*, 2002; Okada, 2006c: 121).

Shifts in the micro-level strategies for survival and the preservation of the seeds for the next-generation technologies led to the modularization of operations, creating a fluid market-like environment and dramatically increasing the decision-making options even to choose partners from competitors. As has been illustrated in this study, one consequence is the development of highly complex macro-level networks for cooperative relations, even suggesting the existence of a 'loosely-networked circle'. This type of development is possible only when network members share a common sense of harmony and cooperation. However, this insight does not at all suggest the rebirth of 'Japan Inc.' given the existence of intense 'modularized competition' at product levels. The emerging picture of networks is so complex that no clear separation between co-operators and competitors is discernible; at least, not by general reference to known company names such as NEC, Hitachi and Toshiba. A company is a cooperating partner in one product area and simultaneously a competitor in another area. Such a complex mixture of cooperation and competition in business relations appears to represent a newly evolving extension of traditional Japanese institutions. However, and as emerging conflicts in some joint ventures suggest, one may be witnessing the development of more complex and effective forms of cooperation less constrained by traditional values and norms.

New Macro-level Strategies

Just as the VLSI Cooperative (1976–79) enabled Japanese semiconductor companies to become world leaders in DRAM technology, once again they appear to have resorted to collective undertakings in order to rebuild their capability. Unlike the past, however, the initial movement did not involve the Japanese government, which feared a revival of the USA-Japan trade disputes. Disappointed by this lack of governmental support,[12] 12 semiconductor producers established (in 1994) a think tank – the Semiconductor Industry Research Institute Japan (SIRIJ) – in order to address problems and needs associated with the development of future technologies and the revival of the Japanese semiconductor manufacturing industry.

One area selected for private firm collaboration was that focused on developing a consortium to rebuild intensive cooperation between semiconductor and equipment manufacturers: jointly, these should design standardized manufacturing processes and develop equipment. Eleven Japanese semiconductor companies established the Semiconductor Leading Edge Technologies or SELETE, (1996–2000).[13]

Unlike the US SEMATECH – which had developed fixed collaborative relations with equipment manufacturers organized as a consortium called SEMI-SEMATECH – the Japanese SELETE operated in a manner rather similar to Toyota's multiple vendor system or Nintendo's software development system, that is adopting an approach more familiar to Japanese managers and one that makes use of competitive pressure among equipment manufacturers. SELETE provided basic specifications to any interested equipment manufacturers in each field, who then competed on the basis of development speed and quality until two winners were selected. A group of engineers from the participating semiconductor producers evaluated equipment prototypes on a test line set up by SELETE, provided criticism and comments on equipment performance, and created an extremely rigorous environment. SELETE made use of the existing Japanese multiple vendor system, introducing market-like competition in vertical cooperative relations. However, it did this with more open, fair and tough relations than were found in the existing Japanese multiple vendor system.

In recent cooperation with the Semiconductor Technology Academic Research Center (STARC) – set up to build universities' semiconductor research and design capabilities – SELETE developed the Advanced System-on-a-Chip through Collaborative Achievement (ASUKA) Project (2001–06) and targeted the development of 100 nanometer (nm) and 70 nm advanced technologies for equipment with multiple products and small-quantity operations (Handotai Sentan Tekunorojizu, 1999).[14]

In contrast to these new forms of cooperation, seven Japanese companies also established a new cooperative, modelled after the old VLSI Cooperative and known as the Super Silicon Crystal Research Institute (SSi, 1996–2001). Its purpose was to produce a 400 mm silicon wafer with a government subsidy. Advancement in wafer technology is vital not only for maintaining leadership in wafer production, but also for reducing the cost of semiconductor production.[15] SSi succeeded in producing a single silicon crystal weighing 438 kg – a worldwide first. Since Japanese wafer producers had never engaged in any collective undertaking, the SSi chief manager feared that competition among them might hinder cooperative activities and waste resources necessary to support the enormously high cost of silicon crystal production experiment (10 million yen per experiment). At the beginning, he could not solicit the participation of company-dispatched researchers in pre-experimental discussions and simulations. However, when he engaged in open discussions based on the result of secret ballots, the wafer producers came to recognize one another's relative strengths and weaknesses and ways to complement each other. This success suggests that competitors are increasingly willing to engage in open, serious, face-to-face discussions and cooperate with each other, although government subsidies clearly remain an indispensable incentive for such cooperation.

Another collective approach was the establishment of the Association of Super-Advanced Electronics Technologies (ASET), again with a government subsidy (1996–2001) to develop (within a period of three years) basic semiconductor technologies for 130 nm line width and (within a period of five years) for 100 nm width.[16] Although the multiple paths for lithography developments increase R&D costs and risks tremendously, sharing basic information among participating members over multiple technologies becomes a very important insurance, since the future advancement of semiconductor technology is highly uncertain. In this context, a governmental subsidy becomes indispensable for reducing risks and costs. By accepting researchers from IBM, TI and Samsung, ASET tried to be open with information and hoped to establish its technologies as future global standards.[17] Openness and information dissemination coupled with good legal protection have become key elements in the development of new strategies – as illustrated by Taplin in this current collection.

In cooperation with the Advanced Semiconductor Research Center (ASRC) at the National Institute of Advanced Industrial Science and Technology (AIST), ASET later developed a virtual organization, the Millennium Research for Advanced Information Technology or MIRAI (2001–07). MIRAI tries to develop new materials and process technologies for semiconductors with 65 to 45 nm line widths.[18]

Although standardizing equipment was attractive to Japanese companies as a means of cutting development costs, some of them found it difficult to enter into agreements with competitors and felt uneasy about cooperating in the development of close-to-the-market equipment. This standardization effort came to require the involvement of the Japanese government, but often with the addition of complex features to justify government subsidies. The first project of this type was the Highly Agile Line Concept Advancement (HALCA, 2001–04), which developed 130 nm system LSI processing equipment with such features as multiple-product, small-lot production and 60 per cent energy efficiency. In a sense, the HALCA project was a sequential stage in developing standardized production equipment following the ASET's earlier lithography development.[19] In contrast, the second project – the Advanced System-on-a-Chip Platform Corporation (ASPLA, 2002–04) – had a clear target of developing 90 nm standardized processing technology and testing new SoC, whose basic technology seems to have been developed by the ASUKA (100–70 nm). ASPLA had a specific goal of overcoming the industry's firm-specific orientation and standardizing process technologies for the system LSI with a government subsidy, in hopes of laying the groundwork for future strategic alliances.[20]

To summarize this depiction of strategic developments, the tradition of collective macro-level strategies with government subsidies served (as it did with the original VLSI Cooperative) drastically to reduce the firm-by-firm costs of developing technologies for the system LSI. While government subsidies were a key mechanism in the past to solicit cooperation among competitors in a consortium, the private-sector initiative undergirding these collective activities also suggests that Japanese companies have become more flexible, open and rational in identifying their mutual interests and cooperating without an external mediator. However, with innovative management practices such as introducing market-oriented competition to replace the old vertical cooperation, legal protection, emphasis on obtaining global standards, secret votes, and face-to-face discussion have become more relevant as elements in successful inter-firm cooperation.

Conclusions: Emerging Patterns of Dynamic Innovation in Japan

Institutions guide or restrict individual and organizational behaviour in certain specific directions and induce repeated patterns of behaviour. Emphasis on cooperation and commitment to long-term relations is an important part of the rules of the game in Japanese business activities. Fitting well with institutions as well as with DRAM technological development, the lifetime employment of workers and cooperative vertical inter-firm relations between semiconductor and equipment manufacturers in the old techno-governance structure became the base of dynamic interaction, cooperative learning and synergetic effects. In the old structure, however, horizontal cooperation among fiercely competing semiconductor manufacturers was possible only in pre-competitive leading-edge research areas and, furthermore, with the help of government subsidies.

This situation changed significantly in the 1990s when Japanese semiconductor companies faced path-disturbing contingencies. Despite the need for drastic and quick changes, top decision-makers feared that unilaterally terminating cooperative relations threatened the established ground for cooperative learning and synergetic effects. Their solution was the introduction of market-oriented mechanisms. For example, a semiconductor manufacturer terminated its long-term relations with equipment manufacturers by telling them to engage in market competition and prove themselves as viable producers, even though there was in fact little prospect of re-establishing cooperative relations.

Similarly, worker lay-offs were more easily accomplished by modularizing operations by creating in-house companies, spinning some of them off as subsidiaries, and selling them away to competitors. Terminating employment by guaranteeing the next employer meant sustaining a company's commitment to its workers' lifetime employment expectation. Other examples illustrated in this current study include modularization, designed to help struggling companies to formulate clear business foci of each separated unit and the corporate headquarters, as well as to provide decision options of selling their spun-off units, buying competitors' units, or forming new alliances. This type of market-oriented approach generated flexibility and fluidity for companies, while it also maintained the Japanese traditional ground for sustaining cooperative learning.

However, and as discussed above, the relatively small size of the modularized units suffered from weaker financial and technological capabilities, rendering them necessary to be divested or merged with or aligned with other companies. Unexpected consequences have included a complex situation of modularized cooperation and competition. One company cooperates closely with another company in one product area, while simultaneously it competes fiercely against the same company in other product areas. Modularized cooperation and competition are a newly emerging and, the author ventures to claim, unique phenomenon of Japanese horizontal-relation-based networks, where cooperation involves nearly all competitors of a given company and occurs whenever actor needs and interests are perceived to match.

Unlike in the past, cooperation between Japanese semiconductor manufacturers is now more open, fluid, calculative and rational; however, it continues to be maintained with some traditional sense of commitment. In order to survive, many traditional Japanese companies need to restabilize their cooperative relations and generate synergetic effects from cooperative learning; and, where necessary, compete fiercely in other product areas. It is this continued emphasis on cooperative learning and synergetic effects – both distinctive Japanese strengths – that keeps the influence of traditional Japanese institutions relevant. It also means that the preferred partners for alliances remain other Japanese companies, since they are assumed to share similar rules of cooperation for synergetic effects.

To rebuild their technological capabilities with limited financial resources, Japanese semiconductor producers revived the traditional system of research cooperative with government subsidies. However, advancing beyond the traditional form, competitors have gathered together to develop various consortia without government subsidies in order jointly to develop equipment and test lines. This suggests a sense of ease in accepting cooperation with competitors even when critically close to market. One implication is that the management of such cooperatives has become much more open, rational, legal-minded and conscious of global standards – without, however, losing a strong emphasis on face-to-face relations.

The long-standing Japanese institution of cooperation thus provides the foundation for new forms of alliances among Japanese semiconductor companies. Though newly emerging conditions shifted the nature of cooperation from vertical to horizontal and that of competition from firm-based to modularized-product based, it has been the enduring value of cooperative learning and synergetic effects that has maintained the continuity of Japanese traditional institutions. Facing path-disturbing contingencies, companies that comprise the Japanese semiconductor industry clearly have developed a new techno-governance structure, but only as an extension of traditional institutions that intricately mix practices of cooperation and competition.

Acknowledgements

The author is grateful to the Murata Science Foundation, the Matsushita International Foundation, and the Grant-in-Aid for Scientific Research (B) by the Society for the Promotion of Science (2000–02) for funding this project. Also, valuable contributions from many interviewees who spared their precious time are deeply appreciated. The author is, however, solely responsible for the content of this paper.

Notes

[1] 'Cooperation' is understood as joint or collaborative behaviour directed toward certain goals based on common interests and mutual expectations.

[2] 'Strategies' are understood as the deliberate creation of defensive capabilities and advantages by allocating and coordinating diverse resources (Goodman & Lawless, 1994: 27).

[3] Information from managers at Toshiba, Sumitomo Electric and Thine Electronics.

[4] Information from a manager at the Electro-Technical Laboratory.

[5] Information from a professor at the University of Tokyo.

[6] Information from managers at NEC, Ando Electric and Hitachi.

[7] Information from a manager at Ando Electric.

[8] Information from a manager at the Institute of Microsystem Integration (IMSI).

[9] Information from managers at Mitsubishi, NEC, Hitachi, Fujitsu Laboratories, Sony and Matsushita.

[10] Information from managers at NEC and Hitachi.

[11] See the Panasonic webpage at http://panasonic.co.jp/corp/news/official.data/data.dir/jn041029-1/jn041029-1.html

[12] Information from managers at Fujitsu Laboratories and Dainippon Screen Manufacturing.

[13] In 2000, it had approximately 170 employees with a ¥10 billion annual budget (Handotai Sentan Tekunorojizu, 1999).

[14] The ASUKA Project had a budget of 76 billion yen for five years jointly financed by SELETE and STARC, with the involvement of 340 staff (see the webpage of the Japan Electronics and Information Technologies Industries Association (JEITA) at http://www.jeita.or.jp/eiaj/japanese/press/pre80.htm) (6 December 2004).

[15] The total SSi budget was 13.3 billion yen, funded jointly by the Japanese government (51.1 per cent) and seven Japanese wafer manufacturers (48.9 per cent).

[16] ASET involved 550 researchers: 200 stationed at a central research laboratory dispatched from participating semiconductor companies and 350 residing at the company laboratories of equipment manufacturers. Its total research budget was about 51 billion yen for five years, taking the form of a research contract with the New Energy and Industrial Technology Development Organization (NEDO), a government organization.

[17] Information from a manager at ASET.

[18] MIRAI involved about 120 researchers at three central research laboratories. A contract with NEDO provided 9.36 billion yen for the first two years (see the MIRAI webpage at http://www.miraipj.jp) (6 October 2004).

[19] HALCA was jointly funded by the government and the companies, totalling 8 billion yen over a three-year period (see the webpage of Semiconductor Portal Co. at http://www.semiconductorportal.com/GSC/shodescr.cfm?nm=HALCA) (6 October 2004).

[20] ASPLA was incorporated with ten companies investing a total of 1.85 yen billion and the government providing a 31.5 billion yen clean room at ASRC in AIST (see the webpage of Semiconductor Portal Co. at http://www.semiconductor portal. com/GSC/shodescr.cfm?nm=ASPLA) (6 October 2004).

References

Casper, S. (2000) Institutional adaptiveness, technology policy, and the diffusion of new business models: the case of German biotechnology, *Organization Studies*, 21(5), pp. 887–914.

Chon, S. (1997) Destroying the myth of vertical integration in the Japanese electronics industry: restructuring in the semiconductor manufacturing equipment industry, *Regional Studies*, 31(1), pp. 25–39.

Daiyamondo (2002) Hitachi dai jigyo saihen (Hitachi makes drastic reorganization), *Daiyamondo* 8 June, pp. 42–48.

DiMaggio, P. J. (1988) Interest and agency in institutional theory, in: W. W. Powell & P. J. DiMaggio (Eds) *Institutional Patterns and Organizations: Culture and Environment*, pp. 3–21 (Cambridge, MA: Ballinger).

Fong, G. R. (1998) Follower at the frontier: international competition and Japanese industrial policy, *International Studies Quarterly*, 42(2), pp. 339–366.

Goodman, R. A. & Lawless, M. W. (1994) *Technology and Strategy: Conceptual Models and Diagnostics* (New York: Oxford University Press).

Hage, J. & Hollingsworth, J. R. (2000) A strategy for the analysis of idea innovation networks and institutions, *Organization Studies*, 21(5), pp. 971–1004.

Hall, R. & Soskice, D. (2001) An introduction to varieties of capitalism, in: R. Hall & D. Soskice (Eds) *Varieties of Capitalism: The Institutional Foundations of Comparative Advantage*, pp. 1–68 (Oxford: Oxford University Press).

Handotai Sangyo Kenkyusho (Semiconductor Industry Research Institute Japan) (2000) *Teigensho: Nihon handotai sangyo no fukkatsu* (Proposal: Revival of the Japanese semiconductor industry), Shinseiki Iinkai (Semiconductor in New Century Committee), March (Tokyo: Handotai Sangyo Kenkyusho).

Handotai Sentan Tekunorojizu (Semiconductor Leading Edge Technologies) (1999) *SELETE 1999*, Annual Report (Tokyo: Handotai Sentan Tekunorojizu).

Lundvall, B.-A. (1988) Innovation as an interactive process: from user-producer interaction to the national system of innovation, in: G. Dosi, C. Freeman, R. Nelson, G. Silverberg & L. Soete (Eds) *Technical Change and Economic Theory*, pp. 349–369 (London: Pinter Publishers).

Lundvall, B.-A. (Ed.) (1992) *National Systems of Innovation: Toward a Theory of Innovation and Interactive Learning* (London: Pinter Publishers).

Lundvall, B.-A. & Maskell, P. (2000) National states and economic development: from national systems of production to national systems of knowledge creation and learning, in: G. L. Clark, M. S. Gertler, M. P. Eldman & K. Williams (Eds) *The Oxford Handbook of Economic Geography*, pp. 353–372 (Oxford: Oxford University Press).

Metcalf, J. S. (1998) *Evolutionary Economics and Creative Destruction* (London: Routledge).

Nagase, N. (2004) *Kaisha-Ho* (Company Law) (Tokyo: Toyo Keizai Shinposha).

Nelson, R. R. (Ed.) (1993) *National Innovation Systems: A Comparative Analysis* (New York: Oxford University Press).

Nelson, R. R. & Winter, S. G. (1977) In search of useful theory of innovation, *Research Policy*, 6(1), pp. 36–76.

Nelson, R. R. & Winter, S. G. (1982) *An Evolutionary Theory of Economic Change* (Cambridge: MA: Harvard University Press).

North, D. (1989) Institutions and economic growth: a historical introduction, *World Development*, 17(9), pp. 1319–1332.

North, D. (1990) *Institutions, Institutional Change, and Economic Performance* (Cambridge: Cambridge University Press).

Okada, Y. (Ed.) (1999) *Japan's Industrial Technology Development: Role of Cooperative Learning and Institutions* (Tokyo: Springer-Verlag).

Okada, Y. (2000) *Competitive-cum-Cooperative Interfirm Relations and Dynamics in the Japanese Semiconductor Industry* (Tokyo: Springer-Verlag).

Okada, Y. (2001) Cooperative learning and Japan's techno-governance structure: exploratory case studies, *Sophia International Review*, 23, pp. 19–42.

Okada, Y. (2006a) Institutions, organizations, and techno-governance for innovation, in: Y. Okada (Ed.) *Struggles for Survival: Institutional and Organizational Changes in Japan's High-Tech Industries*, pp. 9–38 (Tokyo: Springer-Verlag).

Okada, Y. (2006b) Decline of the Japanese semiconductor industry: institutional restrictions and the disintegration of techno-governance, in: Y. Okada (Ed.) *Struggles for Survival: Institutional and Organizational Changes in Japan's High-Tech Industries*, pp. 39–103 (Tokyo: Springer-Verlag).

Okada, Y. (2006c) Institutional changes and corporate strategies for survival in the Japanese semiconductor industry, in: Y. Okada (Ed.) *Struggles for Survival: Institutional and Organizational Changes in Japan's High-Tech Industries*, pp. 105–154 (Tokyo: Springer-Verlag).

Powell, W. W. (1990) Neither market nor hierarchy: network forms of organization, *Research in Organizational Behavior*, 12(1990), pp. 295–336.

Press Journal (1999) *Nihon Handotai Nenkan* (Japan Semiconductor Yearbook) (Tokyo: Press Journal).

Press Journal (2000) *Nihon Handotai Nenkan* (Japan Semiconductor Yearbook) (Tokyo: Press Journal).

Press Journal (2002) *Nihon Handotai Nenkan* (Japan Semiconductor Yearbook) (Tokyo: Press Journal).

Streeck, W. & Yamamura, K. (Eds) (2001) *The Origins of Nonliberal Capitalism* (Ithaca, NY: Cornell University Press).

Toray Keiei Kenkyusho (Toray Management Research Institute) (1997) *Kenkyu kaihatsu shisutemu no seibi ni kansuru kiso chosa: Kenkyu shien taisei no saikochiku ni mukete* (Basic survey on reorganizing the research and development system: rebuilding technology supporting institutions) (Tokyo: Kikai Shinko Kyokai Keizai Kenkyusho).

West, A. (1992) *Innovation Strategy* (Hemel Hempstead: Prentice Hall International).

Williamson, O. (1975) *Markets and Hierarchies: Analysis and Antitrust Implications* (New York: The Free Press).

Williamson, O. (1981) The economics of organization: the transaction cost approach, *American Journal of Sociology*, 87(3), pp. 548–577.

Williamson, O. (1985) *The Economic Institution of Capitalism* (New York: The Free Press).

Yoshikawa, M. (Ed.) (2002) *Kaisha Bunkatsu no Senryaku Katsuyo: Homu, Kaikei, Zeimu no Subete* (Strategies of Company Separation: All about Legal, Accounting, and Tax Procedures) (Tokyo: Zaimu Shohosha).

Innovation, Institutions and Entrepreneurs: The Case of 'Cool Japan'

CORNELIA STORZ

Introduction

The concept of national innovation systems has won recognition in order to explain innovation processes and specific innovation patterns (OECD, 1999). Due to initial coincidences, institutional complementarities and path dependencies, national innovation systems differ permanently in their institutional setting and in the way they stimulate innovation (Lundvall, 1992; Nelson, 1993). The concept thus refers to the fact that specific national institutions give idiosyncratic incentives to actors who generate and implement innovations within the system (Edquist, 1997). Research on national innovation systems – particularly most empirical research on Japan – however, tends to interpret the concept of innovation systems in a very rigid sense. Often the concept is used in order to explain why Japan has lost its competitive position in new industries, such as in business software[1] or in the biotechnology sector (Cottrell, 1996; Goto, 2000). Many authors argue that given the path dependence of innovation systems, Japan is particularly unable to adapt to new technological needs. Its labour market displays low mobility and a low degree of specialization, its capital market has an aversion to risk, its industrial organization prefers long-term transactions and the interlocking between universities and private industries in Japan are weak. All of these characteristics inhibit new industry start-ups (Anchordoguy, 2000). In the 1990s, this discourse caused Japanese officials to implement far-reaching reforms at an astonishingly high speed (Hemmert, 2005, METI, 2001).

Indeed, Japan has a low start-up rate and lacks the existence of Silicon Valley-like small, fast-growing, research-intensive, adventurous firms in new sectors. According to the *Global Entrepreneurship Monitor*, Japan's entrepreneurial activity is the lowest within the Organization for Economic Cooperation and Development (OECD) member states. Linkages between universities and industries, which could induce the foundation of ventures, are practically non-existent (CKC, 2005; GEM, 2005). The focus on reforms revolves mainly around the issue of how Japan can build up a Silicon Valley-like model for venture businesses (Maeda, 2001). This author agrees here that there are indeed areas where reforms of the Japanese innovation system are needed. However, she criticizes a static interpretation of innovation systems. Her argument implies that strict orientation to an 'American model of innovation' may be misleading. On the basis of the internationally successful Japanese game software industry, it is argued that one central factor for this sector's success springs from the unforeseen plasticity of its innovation system.

Hitherto there has been little discussion on the institutional foundations of the Japanese game software sector. Most contributions focus on the political implications of 'soft power' due to the popularity of the industry (Iwabuchi, 2002; JILPT, 2005). Shintaku *et al.* (2004) and Aoyama & Izushi (2003) analyse cultural and structural preconditions for its success, but do not connect their insights to the general question of institutional change and innovation. Using empirical evidence from 30 in-depth interviews with publishers, developers and suppliers of the game software industry and key persons in ministries, associations and research institutes, there will be an analysis here of how and to what extent institutional innovation in the game sector took place. Since the analysis relies on case studies, it possesses exploratory character. Nevertheless, because the industry is made up of a few key firms, the results can claim to be representative of the game software industry.

It is expected that this study will enrich the research on innovation systems. Recently, research on regional breakthrough (Fuchs & Shapira, 2005) and on path creation (Garud and Karnoe, 2001) point in the same direction. Rather than focusing on path dependency, these authors emphasize actors' options to leave paths and/or create new ones and take a more Schumpeterian perspective on how people leave established paths and generate new ones. Since the question as to when a path can be called a new one remains unclear and since the role of already established institutions for innovation tends to be underestimated, it is preferred here to use the term 'plasticity' in conjunction with innovation systems.

Research Question and Definition of Core Research Concept

The dominant institutions of the Japanese innovation system have been identified as being unsupportive for new industries, such as those defining the game software sector. This is a business sector where Japan has become regarded as a global innovation leader – one thinks of global stars such as Nintendo. Against this background, this study addresses the following research question:

● Given the apparent rigidity of many institutional structures and cultures, why has Japan been able to give birth to a new industry, which is so internationally successful: that is, the Japanese game software industry?

In exploring answers to this question, a core research concept has been developed: namely, 'plasticity'. Derived from this concept is the term *plasticid*, a qualitative attribute applied in the concluding comments of this study to describe in part the emerging shape of the Japanese innovation system.

Context: The Nature of Japan's Competitiveness

It is a difficult undertaking to measure the competitiveness of nations. Two indicators that are widely used are a nation's relative comparative advantage as measured by its share of imports and exports (RCA) and a nation's relative share of world trade, which expresses its specialization in export (RWT). Both are not without controversy; for example, neither considers protectionist effects or cyclical fluctuations to a sufficient degree. At most, both measures should be understood solely as rough indicators.

In international trade, Japan has comparative advantages in about 20 commodity groups, almost all of which belong to medium, and not high, technology sectors. Specifically, Japan's exports are in broadcasting, TV, video, optical and photographic equipment, in tool machinery, machinery for certain sectors, and in vehicle construction (Schumacher *et al.*, 2003: 36–40). In more commonly applied terminology, these sectors belong to mechanical engineering and consumer electronics. They possess the common characteristic of having an integral product architecture since their products consist of a high number of single parts whose coordinated development and production is important for the performance of the final product. Further common properties in these sectors are the incrementality of research and development and the importance of process innovation.

At the same time, however, these strengths can be interpreted as comparative disadvantages since they indicate that Japan is not successful in establishing new sectors. Indeed, Japanese firms are almost non-existent in the new software or biotechnology sector. Both of these areas follow more the logic of modular product architectures. The US innovation system, in sharp contrast, was successful in establishing both of these sectors. The fact that these comparative disadvantages existed was not expected by the Japanese side owing to the convincing economic results of the 1980s. For them it came as quite a shock (Kishida & Lynn, 2005) and led to serious discussions about the adaptability of the Japanese innovation system to new economic conditions. Subsequently, it led to a broad consensus on the necessity of reforms (Baba *et al.*, 1995; Hemmert, 2005; *Hitotsubashi Review*, 2005). Outside Japan, as well, the sustainability of the Japanese innovation system became questionable. The OECD stated in one of its policy briefs that an 'upgrading [of] the national innovation system' is necessary (OECD, 2006: 2), while the German Bank declared that 'in Japan, incremental innovation is dominating at the expense of new future sectors' (Deutsche Bank Research, 2006: 11, author's translation).

In other countries, new projects are often taken up by start-ups or spin-offs. In Japan, however, entrepreneurial activity is low. According to the *Global Entrepreneurship Monitor*, only two out of 100 Japanese adults become entrepreneurs, in contrast to the United States (11/100), Canada (9/100) and Germany (5/100) (GEM, 2005). Since the mid-1980s, more enterprises have been liquidated than founded. Between 1999 and 2001, the start-up rate was 3.8 per cent, whereas the rate of liquidations was 4.2 per cent (see Figure 1; CKC, 2005: 229, 396). In absolute numbers, 232,403 firms were founded between 1999 and 2001, while 261,633 failed. These numbers correspond to a negative growth of 29,230 firms.

The low start-up rate becomes even more problematic if one takes a closer look at the motivation of Japanese founders. A high and increasing number of start-ups are founded by relatively older people as a reaction to organizational restructuring. The necessity motive for start-ups is thus much more prevalent in Japan compared to the United States, for example (GEM, 2005: 21). Presently, 30.5 per cent of start-ups in Japan are founded by people in their 50s and 60s, and only 9.9 per cent by persons younger than 29 (KKK, 2006: 5). Empirical research on the relationship between age and innovativeness shows that the age of the entrepreneur is significantly and negatively correlated with the productivity of the start-up (Harada, 2004).

Into this picture fits the insight that venture businesses – understood as independent, research- and risk-oriented, young, and thus mostly small firms that tap into new markets – are rare in Japan. Quantitatively, small and medium-sized enterprises (SMEs) play an important role in the Japanese economy: six million firms belong to this group (which in Japan is defined either by the number of employees (50 to 300) or by the paid-in capital (50 to 300 million yen, depending on the business sector). These SMEs encompass 99.2 per cent of all firms and 79.9 per cent of all employees (CKC, 2005: 382, 386). In an international comparison between OECD countries, only few OECD countries show a higher share of small and medium-sized firms. Very small enterprises with one to 19 employees make up 23.7 per cent of all Japanese enterprises, much higher than in the United States (8.2 per cent) or in Germany (12.4 per cent) (Okamuro, 2002: 42). Nevertheless, supporting a high percentage of small businesses does not automatically translate into innovative activities. Few of these firms can be counted as venture businesses. In 2005, only

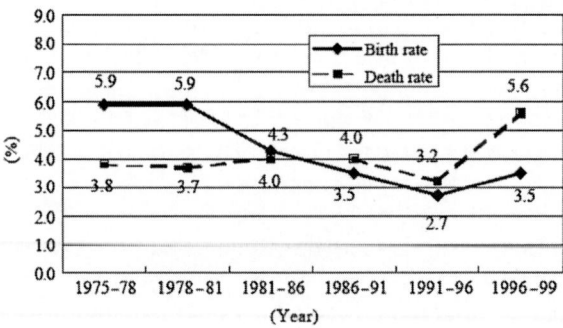

Figure 1. Share of newly established and liquidated firms *Source*: METI (2002: 50)

about 960 firms were listed on the JASDAQ in contrast to the NASDAQ's over 3,600 listed companies (Hwang, 2002; NASDAQ, 2002). Moreover, these few firms belong mostly to established industries, such as machinery, wholesale and retail, and less to new business or technology sectors (Nakagawa, 1999). New sectors are, if at all, founded by affiliated *keiretsu*-firms (enterprise groups), as Müller *et al.* (2004) found for the biotechnology industry. The OECD (2006: 10), therefore, recommended that the 'focus [of] support for R&D [should be] on new start-ups rather than on existing companies as it is now the case'.

How Can the Japanese Innovation System Be Improved?

This question is one of the headings in one of the latest OECD's policy briefs on Japan's innovation system (OECD, 2006). It clearly suggests that reforms in the Japanese innovation system and in Japan's innovation policy are necessary. Simultaneously, Japanese policy makers have initiated comprehensive reforms in innovation-related institutions in order to overcome the dominance of the 'old core technology' (Maeda, 2001: 85).

The concept of national innovation systems (NIS) has been developed by institutional and evolutionary economics in order to explain different national modes of innovation creation and implementation. NIS are defined as a sum of political, economic and social institutions that steer the behaviour of actors into a certain direction (Nelson, 1993; Edquist, 1997). The specific profiles of innovation systems change slowly, resulting in path dependencies and technological paradigms (Dosi, 1982). Specific technological profiles are thus developed by certain institutional configurations. The Japanese innovation system gives different incentives to firms' strategies than the American system does (for the following see Aoyama, 1999; Anchordoguy, 2000; Goto, 2000 – see also Ibata-Arens in this current collection). What follows in this current study is a brief discussion of the 'weak' features of the current Japanese innovation system as seen by many observers.

Weak Features of Japan's NIS

Connecting with analyses presented elsewhere in this current collection, a number of relatively weak features of Japan's NIS have emerged: that is, features that, in global competitive terms, might be improved. These include:

Labour market. The domestic labour market is characterized by long-term tenures and generalized knowledge. The problem of generalized knowledge is directly related to the fact that the education system is geared towards producing generalists rather than specialists. One actual problem is the lack of well-educated software engineers. As new industries need specialized competencies and an influx of external knowledge, the OECD (2006) has recommended more specialization and mobility.[2]

Capital market. Venture capital in the Japanese capital market is rare and is concentrated on established, medium-sized technology industries (Nakagawa, 1999). Firms in new sectors, which due to their innovative character would need risk-friendly seeds capital, have enormous problems in obtaining funds. Entrepreneurs state that the lack of venture capital is the most important barrier to entry (Kawai & Urata, 2002). Moreover, information asymmetries between venture capitalists and investment takers were, until recent reforms, quite high due to the regulation that venture capitalists should not more keep more than 49 per cent of the investment taker's capital. Until 1995, venture capitalists were subjected to restricted regulations with regard to direct and indirect control over investments.

Industrial organization. Japan's industrial organization has traditionally favoured long-term relationships and the enforcement of closed standards. Closed standards work as a brake towards the implementation of global standards, which was necessary in the business software industry. Moreover, the relatively closed character of the standards seems to contribute to a low level of firm entry and to weak domestic competition by lock-in effects, thus reducing incentives to innovate.[3] Finally, innovation management within the firms has built on integral product architectures, which lead to a distinct weakness in giving birth to modular product architectures. The latter are often cited as a prerequisite for new, more open innovation processes (Chesbrough, 1999; McGuire & Dow, 2003).

Interfaces. Innovations not only need appropriate institutions but also appropriate interfaces between institutions. One of the most important interfaces for the generation and implementation of new knowledge is the link between universities and businesses. In Japan, this interlocking is almost non-existent. One historical reason for this lack of working together is the wariness many Japanese still harbour with regard to the pre-war association between research and the military. There are legal and financial reasons as well. Perhaps more striking is the fact that the ratio of patent application out of universities tends to be zero, so that even if informal links do exist, in general the knowing-doing gap in Japan inhibits the creation of ventures. Japan does not lack in ideas (as its high rate in the total application of patents shows), but rather in closing the gap between ideas and their concrete implementation.[4]

The far-reaching reforms in the Japanese innovation system that have been implemented so far aim at radical changes. Networking between public research institutes has been increased by the introduction of temporary contracts. Venture capital has been liberalized by reducing entry and exit barriers for investors via reforms in the tax system and easier terms under which foreign venture capitalists can enter the market. Infrastructural prerequisites and easier access to capital for starting entrepreneurs have been provided, and the transfer of knowledge has been improved by the founding of so-called technology licensing organizations at universities (for more detail, see Hemmert, 2005; Storz, 2005). As outlined in other studies in this current collection, many of these reforms have been inspired by cross-cultural learning, especially from the United States, which serves as a model for Japanese policy makers due to its success with ventures in Silicon Valley.

Competitiveness of Established Innovation Systems: The Japanese Game Software Industry

Parallel to the weak features identified above, a number of distinctive strengths can be deemed to characterize emerging patterns of innovation in Japan. 'Cool' Japanese products are popular among young consumers. Japanese popular culture has been diffused throughout the western world and increasingly questions the dominance of American cultural goods.[5] Despite the fact that Japan has a deficit in business software of about US$3.6 billion, it is a leading producer in the game software subsector (OECD, 1998: 31).

Game software is produced by hardware makers and by specialized game software firms. According to our interviews, the two leading hardware producers, Sony and Nintendo, produce domestically 20 per cent and 80 per cent respectively. The rest of the production is outsourced to software publishers, who develop and produce game software, and to software houses (or developers), who due to resource restraints only develop game software and sell it to publishers or hardware makers for production and marketing. Since developers sell their products often under OEM (Original Equipment Manufacturing), consumers only know the products of publishers. The ten leading publishers ranked according to turnover in 2006 were Sony, Sega, Nintendo, Konami, BandaiNamco, Square Enix, Capcom, Banpresto, Koei and Atlus. Smaller firms, often simultaneously active in other media- or entertainment-related industries, supply specialized elements such as computer graphics, colour design and audio-related projects (compare Table 1).

Approximately 80 per cent of Japanese game software is developed and produced in Tokyo. Since Japanese producers played a dominant role in the world market at least until the 1990s, the district Shibuya in Tokyo, Japan's most important agglomeration of game software firms is an area referred to popularly as 'Bit Valley'. The term signals that Shibuya plays the same role for game software as 'Silicon Valley' does for business software (Baba & Shibuya, 1999). Clusters enable firms to access collective goods without which smaller firms could not balance temporary shortages (Pyke & Sengenberger, 1990). They are also relevant for all knowledge-intensive industries that need the daily contact to consumers to enable 'learning by interacting'. Indeed, the accumulation of game software firms in Tokyo corresponds to the general trend that knowledge-intensive service sectors are based in large cities (Kuratani, 2005). As other metropolises, Tokyo offers a qualified labour pool of graduates from universities and vocational schools together with job changers from firms in related industries. Social networking, the exchange of ideas and the build-up of common knowledge bases are made possible by social events such as the Bit Valley Party. This get-together is attended by hundreds of (mainly young) would-be entrepreneurs, students, software firm representatives, consultants and venture capitalists.

In the development and production of game software, the game software publisher plays the dominant role. Most publishers were founded between the 1950s and the 1970s and employ from a few hundred to a few thousand workers. Thus they fit into the class of what are termed medium-sized firms (in Japanese: *chûken kigyô*). Nevertheless, their size is much smaller than that of the leading firms in the core industries.[6] About half of these firms stepped into the game software sector from related fields, such as entertainment, games or consumer electronics, and most firms were set up 10 to 20 years before launching their first game software (in the case of Nintendo and Sony even longer; see Table 1).

In their properties, these firms stand in contrast to the 'ideal-type' Silicon Valley-model in that they are older, more established firms that have developed game software endogenously at a later point in time. Japanese game software firms are thus a form of 'Japanese-style venture businesses'. They have knowledge intensity and market development in common with venture businesses, but they differ in that they are established firms. This feature is interesting because recent literature suggests that changing technologies necessitate the opening-up of new industries by start-ups or spin-offs.

In contrast to this innovation pattern stands the entrepreneurial activity by game software developers. Presently, there are about 200 to 300 game software developers in Tokyo founded over the last 20 years and from which about 70 per cent have a paid-in capital of less than 50 million yen. According to the Japanese classification system, these belong to the class of small and medium-sized enterprises (SMEs). Most of today's developers were founded isochronously in the 1980s with the launch of new hardware by Nintendo, Sony, or (earlier) Sega (Famicom Mega Drive) and in the 1990s (Playstation, Sega Saturn 1994, Super Famicon; Baba, 1999). The former high entry rates can partially be explained by high profits. Baba (1999: 268) found that under the top 50 firms according to turnover the majority are SMEs. The stagnation of the domestic market, partially due to the upcoming of second-hand game software shops, acts currently as a brake on entrepreneurial dynamism. In our own survey, however, most actors in the Japanese game software sector – be they publishers, developers or smaller suppliers – were optimistic about the future development of the sector and expected an overall positive business climate. One reason may be that the introduction of new platforms by new consoles (Wii of Nintendo and Playstation3 of Sony) and the mobile telephone market offer new opportunities for game software firms.

Game software is played on hardware, primarily consoles, which were almost solely produced by Nintendo and Sony Computer Entertainment (SCE) until the appearance of the X-Box by Microsoft. This is not astonishing due to Japan's comparative advantages in consumer and microelectronics. What is astonishing is the fact that Japanese firms are competitive in game software as well. Owing to an absence of official data – data on software is not compiled separately – one has to rely on private and secondary sources. According to Shintaku *et al.* (2004: 97), for example, the turnover rate of game software more than doubled between 1995 and 1999. About 50 per cent of the game software is exported, a share which has remained stable since 2000 and which contrasts sharply with the high and increasing deficit in business software (DCAJ, 2007: 117). The size of the Japanese market for game software is about 45 billion yen, which is about half of the music market and about 60 per cent of the market for sales of books. After a phase of decreasing turnover (between 1998 to 2003), the market slightly recovered in 2004 (DCAJ, 2005). On the American market, Japanese game software producers nowadays procure about 25 per cent of the titles, while on the Japanese market almost no American firm is present (CESA, 2005: 71). The surplus was about 6 million dollars in 1996, which contrasts with the deficits in the software industry in general (OECD, 1998: 32).

Table 1. Japan's ten leading game software publishers

	Start-up year	Business in start-up year	Launching of game software	Paid-in capital (bio. yen)	Volume of Sales (bio. yen)	Number of employees
SCE[a]	1995	Consumer electronics	1996	1.93	957.20	1200
Sega[b]	1951	Entertainment games	1983	26.00	533.24	3050
Nintendo	1989	Game cards	1977	10.06	298.81	4500
Konami	1969	Entertainment games	1978	47.30	262.10	5000
BandaiNamco[c]	1950	Game equipment for department houses, driving parks, arcade games	1974	10.00	206.63	6776
SquareEnix[d]	1975	Research on public house building	1986	7.80	124.40	3050
Capcom	1979	Electronic games, arcade games	1983	27.50	70.25	1200
Banpresto[e]	1977	Manufacturing und sales for arcade video game machines	1990	3.02	34.43	184
Koei	1992	Business software	1981	9.09	26.22	725
Atlus	1986	Computer games	1989	8.45	16.73	324

Source: Compiled from official reports on the firms' websites[7]
Note: [a]Both hardware and software is only developed by SCE (Sony Computer Entertainment), not by Sony.
[b]In 2004 fused with Sammy into SegaSammy (general entertainment).
[c]Namco was founded in 1974 and fused tc BandaiNamco in 2006.
[d]Fused to SquareEnix in 2003. Square was founded in 1986 in the software development sector and is the producer of Final Fantasy (1987).
[e]Cooperates with Mitsubishi.

The Plasticity of Innovation Systems

Developing on the concept of 'plasticity' in relation to national innovation systems, it is possible to explain how Japanese entrepreneurs have created their own innovation system for game software.

Institutional Foundations for Plasticity

The plasticity of innovation systems stems from the institutional and structural variation of given systems. The concept of path dependence starts from the assumption that institutions are complementary and mutually reinforcing. In reality, however, innovation systems are made up of a set of institutions that are dominant or peripheral in character (Amable, 2003). Consequently, different institutional solutions tend to exist for different innovation problems. Thus, actors do not just follow the incentive of a given, monolithic innovation system. Rather, they select those institutions that fit their needs out of the institutional pool (as more recent research on regional and sectoral innovations argues as well (Malerba, 2006)).

Based on the notion of variation within given systems, Streeck & Thelen (2005) have developed five modes of institutional change. In this study, two of them are employed: conversion and displacement. The conversion of institutions means that institutions are redirected to new purposes – in our case the purpose of developing game software. The important point here is that by way of conversion, even those institutions that seem to have hindered innovation in the past may favour innovation in other contexts. If one follows this approach, then the assumption of self-enforcing institutions (as we know it from the concept of path dependence) becomes questionable. Self-enforcement takes place outside of actors, whereas conversion presumes creative actors converting institutions to new purposes. A second source of plasticity is displacement, which refers to the fact that new models raise questions about established institutional forms. These new models may draw on peripheral institutions or combine with converted, established institutions. A further form of displacement is the creation of wholly new institutions. Both may undermine the dominant system over a longer period. We now look more closely at the two modes of 'conversion' and 'displacement'.

Conversion

The Japanese labour and capital markets as well as Japan's industrial organization have been identified as barriers to the genesis of new industries. In the following discussion, however, it becomes clear that not all 'future industries' necessitate the same institutional setting. Even 'old' institutions may be appropriate when they are converted to new purposes or when they are combined with new elements.

Japanese firms possess the distinct ability to activate implicit knowledge and to transfer it into explicit knowledge. Several authors have advanced the thesis that the 'Japanese style of management,' – long-term employment – leads to more open communication flows which may enable firms to solve problems of heterogeneity, uncertainty and ambiguity better (Aoki, 1992; Nonaka & Takeuchi,

1995). The important point in this context is that the development of game software is confronted by these problems to an even greater degree than the dominant core industries are.

Heterogeneity. The heterogeneity of participating disciplines in the development of game software is high and ranges from artists (game designers, illustrators and sound engineers) to engineers (programmers and testers),[8] and even to machinery engineers in the case of cooperation with hardware makers. The different knowledge bases make knowledge sharing difficult.

Uncertainty. Closely related to the problem of different knowledge bases is the problem of uncertainty. Uncertainty results from the fact that artistic ideas have to be translated into computer language, but it is not at all clear beforehand whether or not and to what extent these ideas can be transferred into technical language. Furthermore, the innovation process itself may be more uncertain than that of the dominant core industries. For example, a car design does not change with the computer background, but the characteristics of figures may change through their movements or through different degrees of brightness on the background, as game developers have often reported.

Ambiguity. The internal coherence of a game is essential for a game's success. Artists and engineers need to develop a common idea of the game and of the single moves in the game. The first ideas of a game, as well as its very first drafts, are ambiguous and difficult to communicate since large parts rely on personal knowledge and ideas. This is even the case for the whole development process. In the very beginning of a game there are only unanimated cartoons. Their animation is realized by programmers who have to understand the game designer's (and his core group's) concept of what the figure should look like, how the figure should move, which facial expressions it should have, how realistic it should be, and so on. Whether game designers or illustrators can receive more 'exciting' or 'more floating' looking figures, to take just a simple example, depends on the programmer's implicit understanding. He or she has to translate that expression into a concrete program. Another example is the testing phase. Testers may state that the game should make a 'more coherent' or 'brighter' image. Here, again, interpretation and implementation need implicit understanding. In the automobile industry, one can (with relative reliability) predict how many cars will be sold given a certain quality, price and design. In game software, sales are highly volatile and depend merely on the consumers' tastes, that is, whether they find the product 'exciting' or 'cool'. Approval, in turn, depends on creative designs and on their coherent implementation.

Given this typology of requirements it may be no coincidence that Japanese publishers and larger developers tend to rely on the 'J-model' of management, especially on a relatively low level of staff mobility and on generalists. Both aspects improve internal knowledge flows and induce a shared yet varied knowledge base. For the most part, the creative staff as well as programmers and testers in Japanese firms are employed as *shain*s, that is, as long-term employees. One software publisher reported that artists are employed on a project basis in order to stimulate the inflow of new external knowledge and to have more flexibility in lay-offs in the case of unsatisfactory creative abilities. In reality, however, artists in this firm are employed only at first on a temporary basis. Even during this preliminary period, they often work in the firm, and about half of them are taken over months after the project has been completed. In other words, an institutional innovation does take place in that new labour markets are becoming open, but this effect is only a temporary one.

Although at first glance it may appear that more mobile solutions are being employed, in actuality it is more the case that established institutions are being reinterpreted. Another mechanism that is used by many firms is to employ students as temporary hires (*arubaito*). Here, again, it is often the case that these students are later employed by the firm. In both cases, then, the concept of long-term employment is retained. Our interview partners showed a clear preference for the creation of a common knowledge base and the openness of communication at the expense of external inflows of knowledge. Sunagawa (1997) reports as well, drawing on the case of Sega, that on-the-job training is used in order to learn by mistakes and to improve the transfer of tacit knowledge.

In terms of skill development, the instrument of job rotation, which is widely established in the Japanese industrial core, is applied with the same goal: to create open horizontal communication flows. Moreover, it serves as an instrument of job enrichment. Originally, job rotation was applied to excellent game designers who were urged by management to develop (commercially more attractive) series. (Presently, nine out of the ten most successful products are series.) This instrument was later commonly used for employees involved in game production, since the development and production of series are not very demanding, even boring from an artist's point of view due to their stable framework.

An underdeveloped capital market was also identified as a factor for Japan's weakness in new industries. As game software development does not need very much capital in the beginning and can generally be financed internally, lack of funds is not a major problem at the start of the industry. Instead of institutional venture capital, start-ups were often sponsored by quasi-private venture capital. The publisher NamcoBandai reported that it supported more than ten developers in the 1980s, and that this type of support was quite common in the game software industry. Sega supports employees who want to become independent entrepreneurs with its 'programme for start-ups' (*dokuritsu shien program*). In addition to capital, the programme provides services as well, such as legal advice, advice related to intellectual property rights and technical support.

To date, Sega has sponsored nine firms. This type of angel financing is quite interesting since it has its parallels in the already established institution of sponsored spin-offs – temporary investments by an employer of a small or medium-sized firm in the start-up of a former employee. There are several incentives for game publishers to function as a sponsor. Firstly, start-ups are under more pressure to get their products ready for marketing than established firms. Sponsoring thus fosters 'economies of speed'. However, this advantage may be undermined if the sponsored founder behaves opportunistically by pursuing opposing goals to that of the incubator. Sponsoring is thus an instrument that can be employed to secure congruency between the sponsoring and the sponsored firms' strategies, as well as to help create a favourable basis for coordination. This includes creating an atmosphere in which former employees are not unduly discouraged when they are interested in becoming entrepreneurs. Instead of this, the former in-house career is supplemented by an out-of-house career in the new start-up.

Furthermore, outsourcing certain competences is one way to concentrate on one's own core competencies (see Storz & Frick, 1999). In the case of the game software sector, the lack of venture capital may thus even be interpreted as a comparative advantage in that sponsoring publishers were able to provide highly qualified and specialized in-house consulting. Providing start-up capital for the seeds-phase by business angels is thus not new in the Japanese context. Its purpose is similar to the traditional form of sponsoring: to ease coordination and to motivate former employees. What is new is that it is being applied in a knowledge-intensive sector and to new actors, and, due to its transfer into a new sector, to the purpose of sponsoring itself. Now, series production is partially outsourced so that internal creators have more time and energy to create fascinating, new, crazy games.[9] Furthermore, sponsoring encourages the timely coordination of game development, which is important due to games' short life cycles (two to three months).

Finally, industrial organization has been identified as a barrier for the development of new industries, especially for the software industry. One argument is that the focus on long-term co-operation restricts the influx of external knowledge, inter-firm competition and the setting up of globally competitive standards (McGuire & Dow, 2003). Long-term cooperation, however, makes the creation of instruments possible, which again facilitates knowledge sharing such as the intra-firm exchange of engineers and designers from developers to publishers and vice versa. Interview partners in the game software industry identified the sharing of knowledge as one of the most important instruments for speeding up the innovation process and reducing software bugs and incoherent storylines. The fact that Japanese firms are much more present in role playing games on the world market than American producers are may be explained by intense inter-firm cooperation, which is the precondition for creating complex games that need a high degree of coordination.

Displacement

Besides existing institutions, new institutions in the Japanese labour market and industrial organization have emerged. Many of them have been selected from the periphery and have led to changes in dominant institutions as well; others have been newly created.

In labour market terms, the game software sector needs a highly specialized workforce. At the very start of the industry, no specialized education profiles for game software existed and its market size was still quite small. There were enough programmers to go around, and the game industry was often supplied by a group who were dissatisfied with their work in other (often defence-related) software sectors and wanted a change. The relatively small size of the business software industry was in this case a favouring factor in that it did not absorb the scarce resources of qualified human capital. The technological infrastructure of games, however, has become more and more complex so that the question of specialization has also become a major issue. Today, switching back and forth between different software sectors is no longer possible. Firms producing business software components reported in our interviews that while in the beginning they served both business and game software firms simultaneously, presently specialization in one or the other software segment is necessary.

Thus, the most general problem the Japanese labour market faces today is that the degree of specialization offered by educational organizations is relatively low. One result is a shortage of people who are both qualified in design and technical software knowledge. This lacuna has opened up opportunities for private entrepreneurs and led to entirely new institutions. Private education institutes teaching courses that integrate design, computer graphics, programming, engineering and so on have sprung up all over Japan. Presently there are estimated to be 109 new schools related to the digital media industry, about 30 per cent of them founded by private entrepreneurs (Baba & Shibuya, 2000: 38). One of the best-known institutes is the Digital Hollywood University. It was founded in 1994 and has since established ten branches. Digital Hollywood offers classes early in the morning and late in the evening so that working people can attend as well. Between 1994 and 2006, 36,000 people graduated. According to the school's director and one external key informant, about 80 per cent of graduates have found a place in the entertainment and digital media industry.

A further institutional innovation is the establishment of networking associations. The non-profit organization BVA (Bit Valley Association) was set up in 1999 and aims at promoting personal contacts between people engaged in Internet business. Their social events, such as the Bit Valley parties, are important instruments not only for fostering mutual contacts between game-interested persons[10] but also for finding capable personnel (see Yukawa, 2003: 3). They provide another reason why Tokyo is the most attractive location for game-related businesses.

Within the firms, the new requirements in the game software sector with regard to creativity and individuality as well as to the personal background of the founders, who understand themselves as artists with a 'non-mainstream' personality, brought about new management methods. The dominant core industries have become more important sources of personnel than the universities. This fact is interesting since it contradicts the dominant logic in personnel management. Generally, a majority of new employees were recruited from universities, since individual creativity was not thought to play a large role in the industrial core technologies. Certainly, in the first stage of selection common criteria, such as literacy or general knowledge, stand out. In further steps, however, individual creativity and an 'artistic personality' are the most important selection criteria for the game industry. The creativity of applicants is identified by personal interviews in which applicants have to solve certain problems, such as design certain parts of a game or invent an interesting idea for a game based on certain given facts. Thus, somewhat 'crazy' but sufficiently qualified students have a relatively good chance of finding a place in game software firms.

In terms of intra-firm organization, the volatility of the market was encountered by an abandonment of strategic investments. Hardware makers in the game software sector – Sony, Nintendo and the former Sega – have always refrained from investing in publishers, even if transactions are long term. Instead, publishers and developers pay an *ex ante* fixed royalty according to the required ROM (read only memory) and the size of order (Yanagawa, 2004). Their cooperation is thus based on agreements termed by Kohashi and Kagono (1995) a 'rule-regulated network' and which are quite similar to project networks, as we know them from the television and movie industry. They are an organizational form somewhat between legally independent but economically dependent firms for the realization of temporarily limited, complex and innovative projects. This kind of organization stands in sharp contrast to the industrial organization in the dominant core industries, in which makers such as Toyota or Nissan hold about 30 per cent of their suppliers' capital (see Storz, 2006a, 2008).

Discussion

At first glance, the evident success of the Japanese game software industry contradicts much research focused on arguing that certain institutions have worked as barriers towards the genesis of new industries in Japan, and especially in respect of the software industry. It has been argued here that Japan's successful game software sector can be explained by the plasticity of innovation systems. Another approach could be that those institutions that hindered the genesis of Japan's business software industry are meaningless for its game software industry. In other words, one can argue that we do not know the range of industries that are compatible with a given national innovation system, so that there always exist untapped, new, and yet to be discovered industries that are compatible with existing innovation systems.

Applying this insight to the Japanese innovation system, we can see that a multitude of compatible industries may exist. Japan's economic success in so many industries could be taken to mean that the Japanese model works fine in general, and that, for whatever reason, business software and biotechnology are the anomalies. At the very least we can hypothesize that numerous 'other' future industries exist that fit with the institutional design of the Japanese innovation system. There are findings, for example, that suggest that Japan is highly competitive in nanotechnology (Marinova & McAleer, 2003),[11] in wireless Internet and in i-mode. The success of these industries has been traced back to their compatibility with the Japanese innovation system (Ratliff, 2004). From this perspective, the genesis of the game software industry can be understood as a 'logical consequence' of given institutions and structures. What this approach neglects, however, is that industries do not just 'fit in' to established institutions, but that institutional adaptations and innovations are also necessary in order to create a 'fitting' institutional environment. In other words, there may be cases where a fit is given; however, in many other cases this fit has to be created.

In the case of the game software sector, established institutions from the industrial core sectors had to be introduced into the game software industry. In the process, these established institutions became transformed by peripheral institutions. For example, long-established management methods in the domestic labour market were maintained, but they were combined with new methods in personnel assignment and recruitment. In industrial organization, traditional instruments for knowledge exchange were kept at the downstream firms (publishers and developers), but the publishers themselves became more independent due to the lack of strategic investments by hardware makers. Institutions are thus neither given nor static, and a new set of appropriate institutions can be created. Even if we still have to identify the conditions under which actors convert and displace institutions, this is a creative act and a prerequisite for the plasticity of innovation systems. Plasticity is a different concept from a static, actor-less concept of 'fitting'.[12]

At the same time, it is also true that non-institutional factors favoured the development of the game software industry. While an in-depth analysis goes well beyond the scope of this paper, a few important factors – especially relevant to explaining the genesis of the game software sector – can be listed. These are cultural knowledge stocks, a strong demand for game software and competitive electronic producers. First, the tradition of *manga*, a certain cartoon-style expanding a history of more than 80 years, indicates that Japanese designers and artists possessed a unique stock of knowledge. Second, while the total volume of the entertainment and media industry in Japan is about 13 billion yen and thus lower than that for the United States (34 billion yen with only 2.2 times the number of inhabitants), the market for game software is in both countries about the same (Japan: 1 billion yen; USA: 0.9 billion yen – see DCAJ, 2005). Third, the given industrial structure was an important factor for the genesis of the industry. The strong position of hardware makers in consumer and microelectronics helped game software firms build upon existing networks in production, marketing and distribution. This made closing the 'knowing-doing gap' easier in that product developments could be commercialized relatively smoothly. Due to the knowledge base in microelectronics, the memory capacity of the hardware was quite high from the beginning, which again simplifies the production of attractive design by visualization and the use of new techniques. According to the concept of lead markets (see Beise, 2004), the ability of an innovation design to diffuse internationally is positively correlated with the attributes of the country that first developed that design, especially with regard to demand and market structure.

Up to this point, the discussion has focused on the institutional and structural foundations that have given the Japanese game software industry a structural advantage. Owing to Japan's relative global weakness in the software industry in general, the comparative advantage in game software is in need of an explanation. However, Japan's output of game software shipments declined from 2000 on and only slowly recovered in 2004. This shift was caused by shrinking demand inside and outside Japan, while at the same time foreign markets for game software grew.[13] Japanese firms did not participate in the growing markets. They basically left the field to competitors such that the share of Japanese game software in the United States decreased from its former 50 per cent to its present 25 per cent (DCAJ, 2007: 117). Even if these data can be qualified by shedding light on other empirical realities, such as the fact that the traditionally obtained software industry shipments do not include monthly membership fees for players participating and subscribing to online games (Eurotechnology, 2005: 45), the overall impression is that the Japanese game software industry has lost some of its dynamic.

So far there is no clear analysis as to why this should be the case. One reason suggested by one of my interview partners is that Japan has not been able systematically to integrate external knowledge. In this study it has been argued that the Japanese innovation management inside and between firms generated comparative advantages for developers in that it made the transfer of tacit knowledge easier. Increasing competitive pressure, shortening product cycles and increasing development costs, however, seem to force firms to change their innovation management from formerly closed to more open innovation processes. The traditional focus on core competencies generating inter-organizational learning, which was considered hitherto a comparative advantage over competitors who were less able to activate implicit knowledge and to transfer it beyond firm boundaries, may be too narrow a focus for explaining innovation in Japan today.

A focus on the 'knowledge creating company' (Nonaka & Takeuchi, 1995) neglects a perhaps much more important source of innovation: the knowledge created from outside the firm (Chesbrough, 2003; Jeppesen & Molin, 2003). As a result, learning by searching – an institutionalized, formalized process, oriented towards a certain innovation – loses its importance for many new industries, while learning by experimentation or 'creative improvisation' – as Schrage (1999) called it – becomes decisive for the creation of new products (Sundbo, 2000). Chesbrough (2003, 2006) convincingly argues for a new model of open innovation processes by which firms make use of external as well as internal ideas. In this model, the locus of innovation shifts from the in-house laboratory to consumers, universities, start-ups, research consortia and other organizations. As a precursor in this respect, von Hippel (1986) shifted our attention to the concept of user-driven innovation and Allen (1983) to collective invention, but changing product architectures and new technologies such as computer-based simulation or modelling (Thomke, 2003; Dodgson *et al.*, 2007) have increased the options of extending the learning to an outside community, thus offering new potential for the creation of innovation.

This is not to argue that all industries will migrate to open innovation processes. In game software, user-centred design methods have 'yet to find their way into the ... industry' as a Microsoft Corporation research report states (Pagulayan *et al.*, 2003). Nevertheless, there are several signs that many industries are in a transition phase. With Open Source Software, current software development belongs to the most prominent examples of open innovation. At the same time, recent case studies point to the increasing role of consumer communities in innovation processes in game software as well (Jeppesen & Molin, 2003; Haddon *et al.*, 2005; Prügl & Schreier, 2006), and new forms of cooperation between game developers, universities and other organizations can be observed. Specialized conferences – such as the Game Developers Conference in the United States, at which reasons for the competitiveness of games are analysed on a high level and in a systematic way – take place on a regular basis.

For the most part, however, and according to my interview partners, shifting the locus of innovation takes place only peripherally in the Japanese game software industry. Certainly, conferences, industry fairs or informal networks as that described by the BVA example above are also important instruments for knowledge creation in Japan. Nevertheless, it seems that a more systematic and demanding analysis in close contact with external knowledge sources plays a subordinate role. Not one of the interview partners mentioned that he or she has dense contacts to universities, analysts or other external experts. Furthermore, venture capitalists also play a minor role. The ventures described in this study are mostly internally financed, and many of them are in fact part of the company's group. It thus seems that innovation processes in the game software sector remain quite closed.

Consequently, one cannot prove here whether there is indeed a causality between not yet established open innovation processes and decreasing market shares, nor can one go deeper into the question of whether or not or to what degree innovation processes in the game software sector will transform into open innovation processes. However, if open innovation indeed becomes a 'new paradigm for corporate innovation' (West & Gallagher, 2006), Japanese firms have to respond quickly. Their presence in the United States and in the leading American game software association ESA (Entertainment Software Association), as well as the often-cited ability to learn and to adapt, might help in this process. Another indicator for Japan's transition may be the fact that the Digital Games Research Association was founded in Japan in 2006 and follows the US model of the Game Developers Conference by trying systematically to include knowledge bases outside the firm (Storz, 2006b).

Conclusions and Implications for the Japanese Innovation

The comparative advantage that Japanese firms possess in game software is surprising. At first glance it appears to contradict institutional theory, which lays the blame for Japan's weakness in the software industry on institutions such as the country's underdeveloped labour or capital markets. A superficial explanation for the development of a game software industry could draw on the 'fitting' of the game software sector to the existing institutional setting. From this perspective, game software would simply need institutions and interfaces other than those of the business software sector. To some degree this approach may be valid. However, such a theory neglects that the fact that what fits does not necessarily exist but must be created by entrepreneurs. This process is exactly what happened in the Japanese game software industry: entrepreneurs created their own innovation system for game software by converting and displacing institutions (Streeck & Thelen, 2005) and by combining them to form a new institutional setting. The Japanese example shows that innovation systems possess an important property, defined in the context of this study as *plasticid.*

Conversion and displacement are two types of institutional change. Conversion means that established institutions are channelled to fit a new purpose. In the case of Japan's game software industry, established institutions, such as institutions of knowledge creation, were transferred into the game software sector, thus contributing to the solution of problems of heterogeneity (of actors' knowledge), uncertainty and ambiguity (of projects). Even institutions that have been identified as being inappropriate for new needs appeared to be appropriate and supportive to these needs once channelled to different purposes.

Displacement is another factor for an innovation system's plasticity. Recent publications on the diversity of innovation systems notwithstanding (Malerba &

Orsenigo, 1996; Lundvall *et al.*, 2002), the literature on innovation systems often refers only to dominant institutions in explaining why certain nations were unsuccessful in establishing new sectors. In reality, however, numerous, alternative, peripheral institutions exist. This is exactly what one sees in the game software sector in Japan: peripheral institutions of the labour market – such as the selection of personnel on the basis of their creativity and individuality – have been important prerequisites for the genesis of the game software industry. Firms sought out and attracted young, performance-oriented designers, thereby displacing established institutions in personnel management.

In terms of broader policy implications, the case of the Japanese game software sector teaches us that politically it may be wise to embrace variety. The variation of a given set of dominant and peripheral institutions opens up options for new combinations. Moreover, those institutions that were identified as having hindered a competitive edge for the business software sector were – at least partially – exactly those institutions that induced Japan's game software industry to be so competitive. It should not have to be pointed out that in a concrete situation it is difficult to decide which degree of variety should be maintained and which institutions should not become the objects of a paradigm change within a given innovation system. What one can take from the Japanese case is that a lack of complementarity need not act as a barrier to institutional change; at least not in the case of the game software sector, where the lack of complementarities was not perceived as a difficulty by Japanese entrepreneurs. On the contrary, they perceived that institutions from the dominant sector could be adapted to changing needs in conjunction with peripheral institutions.

Finally, the Japanese case shows that there is no one 'best practice' model for innovation, and that innovation need not occur through start-ups in the form of a Silicon Valley model. With few exceptions, the leading Japanese game software publishers are not recent start-ups. Most of them were founded in the decades from 1950 to 1970 and started the game software sector drawing on their experience in other business segments. The case of the Japanese game software sector shows that an institutional setting such as an innovation system is (paraphrasing from Nelson & Sampat, 2001) 'like a paved road across a swamp'. To say that the location of the prevailing road is a 'constraint' on getting across is, basically, to miss the point. Without a road, getting across would be impossible. When actors are creative and innovative, plasticity may work to counteract institutional rigidity.

Scope for Further Research

Having said this, some qualifications should be made. First of all, the results of this study are based on explorative case studies. While they have indeed included many leading game software publishers, they have not by far incorporated all relevant actors. Moreover, the question has not been answered here as to which element – the conversion of established institutions or displacement by new institutions – has been more important for the genesis of the industry. Nor is it possible to determine the extent to which the distinct Japanese innovation pattern with a relatively strong focus on in-house and between firms dialogue rather than on external knowledge, for example, may become a barrier to further competitiveness. Even if there is no one optimal innovation process, a few adjustments in Japan may be necessary. In the face of declining market shares in the United States, these questions are all highly relevant.

Finally, the question has not been answered here concerning under which conditions actors are able to convert and displace institutions, nor were non-institutional factors sufficiently integrated here, such as existing knowledge (for example, the *manga* industry), the established industrial structure and the demand for digital entertainment goods. Further research needs to address these questions. These open questions notwithstanding, the case of the game software industry is an instructive example of how entrepreneurial a country Japan is, even if its patterns of entrepreneurial activity differ.

Acknowledgements

Research was conducted in Japan in May and June 2006 and was made possible due to an invitation of the Japanese Institute for Labour Policy and Training (JILPT). The author thanks the Institute for the generous support. The author offers thanks for valuable comments at workshops within the JILPT, as well as thanking the members of the Standing Working Group on Comparative Study of Economic Organization at the 21st EGOS Colloquium, Berlin, Germany, especially Glenn Morgan. Special thanks go to two anonymous referees and their valuable hints. Further the author would like to thank especially some of my interview partners in Japan: Kazuhiko Abe, Shun Arai, Akira Baba, Takumi Hirai, Mari Kawai, Tomikazu Kirita, Shintaro Kobayashi, Akira Kudo, Shigeru Matsushima, Michihiro Sasaki, Junjiro Shintaku, Hirohide Sugiura, Noriyuki Takahashi, Hajime Wakuda, Hajime Yamada, Masaru Yamashita, Noriyuki Yanagawa and Nobuyuki Yoshizawa. Finally, the author thanks Marcel Loginow and Caroline Siber for their research assistance.

Notes

[1] Baba *et al.* (1996: 114) classify software products into finance, operating systems, office automation, science and technology, telecommunications, defence and games. All categories except 'game software' are referred to here as 'business software'.

[2] It is true that listed companies have low mobility rates. The labour market between smaller firms, however, is much more permeable. In small firms with 5–29 employees, 70.3% of new appointments are changers (*tenshoku*) in contrast to firms with more than 1,000 employees (52.2%; see CKC, 2005: 196).

[3] Kawai and Urata (2002) come to different results. In their investigation on the determinants of entry, they find that subcontracting opportunities promote entry. They suggest that the Japanese subcontracting system is open to newcomers.

[4] Besides unfavourable institutional conditions, the stagnation of the Japanese economy from 1992 on was an important barrier for market entries (Harada, 2005).

[5] Cultural goods are classified as the 'entertainment and media industry'. Its subsectors are filmed entertainment, television networks, television distribution, recorded music, radio, Internet advertising, business information, magazines, newspapers, consumer books, educational books and training publishing, theme parks, sports and video games (Chûô Aoyama, 2005).

[6] Toyota, for example, counts 265,000 employees.

[7] In the Appendix, the size of Japan's leading publishers is compared to the size of US firms.

[8] This does not mean that programmers are not creative.

[9] Depending on the firm's strategy, another option is to outsource more risky, new games. Sponsored spin-offs are then an instrument of risk optimization.

[10] Most of the employees in the game sector play themselves as well. They thus form a large 'game community' (*otaku*).

[11] Japan has the world's highest rate for assigned patents in nanotechnology (Marinova and McAleer, 2003).

[12] From this perspective one may also question the extent to which dominant institutions hindered the genesis of the business software industry. The requirements of both sectors on both the labour market and the capital market are quite similar. Network externalities and the strategic decision of Japanese producers not to open up software standards seem to be much more convincing reasons than the inhibiting argument. The causality that has been drawn between 'institutions' and 'lack of incentives towards the software sector' seems to be weaker than originally thought.

[13] The volume of game software on the US market doubled between 2001 and 2006, and the volume on the European market tripled (METI, 2006).

References

Allen, R. C. (1983) Collective invention, *Journal of Economic Behavior and Organization*, 4(1), pp. 1–24.

Amable, B. (2003) *The Diversity of Modern Capitalism* (Oxford: Oxford University Press).

Anchordoguy, M. (2000) Japan's software industry: a failure of institutions, *Research Policy*, 29(3), pp. 391–408.

Aoki, M. (1992) Decentralization-cetralization in Japanese organization: a duality principle. In. *Japanese Political Economy Vol. 3* (Palo Alto: Staford University Press), pp. 142–169.

Aoyama, Y. (1999) Historical underpinnings of the small business policy in Japan and the United States, *Small Business Economics*, 12(3), pp. 217–231.

Aoyama, Y. & Izushi, H. (2003) Hardware gimmick or cultural innovation? Technological, cultural, and social foundations of the Japanese video game industry, *Research Policy*, 32(3), pp. 423–444.

Baba, Y. & Shibuya, M. (1999) Tôkyô gêmu sofuto kurasutâ (Tokyo's games software cluster), *Kenkyû Gijutsu Keikaku*, 14(4), pp. 266–278.

Baba, Y. & Shibuya, M. (2000) Tôkyô gêmu sofuto kurasutâ (Game software cluster in Tokyo), *Kenkyû Gijutsu Keikaku*, 15(1), pp. 33–47 (in Japanese).

Baba, Y., Takai, S. & Mizuta, Y. (1995) The Japanese software industry: the hub structure approach, *Research Policy*, 24(3), pp. 473–486.

Baba, Y., Takai, S. & Mizuta, Y. (1996) The user-driven evolution of the Japanese software industry: the case of customized software for mainframes, in: D. C. Mowery (Ed.) *The International Computer Software Industry*, pp. 104–130 (Oxford: Oxford University Press).

Beise, M. (2004) Lead markets, drivers of the global diffusion of innovations, Kobe University Discussion Paper (RIEB), Available at http://www.rieb.kobe-u.ac.jp/academic/ra/dp/English/dp141.pdf (accessed 12 December 2006).

CESA (Computer Entertainment Supplier's Association) (2005) *2005 CESA Games White Paper* (Tokyo: CESA).

Chesbrough, H. W. (1999) The organizational impact of technological change: a comparative theory of national institutional factors, *Industrial and Corporate Change*, 8(3), pp. 447–485.

Chesbrough, H. W. (2003) The era of open innovation, in *MIT Sloan Management, Review*, 44(3), pp. 35–41.

Chesbrough, H. W. (Ed.) (2006) *Open innovation: researching a new paradigm* (Oxford: Oxford University Press).

Chûô Aoyama (Ed.) (2005) *Kansahôjin kontentsu bijinesu handobukku* (Handbook of Entertainment and Media Industry) (Tokyo: Chûô Keizaisha) (in Japanese).

CKC (Chûshô kigyôchô) (2005) *Chûshô Kigyô Hakusho* (White Book on KMUs (SMEs)) (Tokyo: CKC).

Cottrell, T. (1996) Standards and the arrested development of Japan's microcomputer software industry, in: D. C. Mowery (Ed.) *The International Computer Software Industry*, pp. 131–164 (Oxford: Oxford University Press).

DCAJ (Digital Contents Association Japan) (2005) *Dejitaru Kontentsu Hakusho* (White Book on Digital Contents Industries) (Tokyo: Digital Contents Association Japan).

DCAJ (Digital Contents Association Japan) (2007) *Dejitaru Kontentsu Hakusho* (White Book on Digital Contents Industries) (Tokyo).

Deutsche Bank Research (2006) Japan 2020 – ein steiniger Weg (Japan – a rocky road ahead) Available at http://www.dbresearch.de/PROD/DBR_INTERNET_DE-ROD/PROD000000000020265 (accessed 18 September 2006) (in German).

Dodgson, M., Gann, D. & Salter, A. (2007) *Think, Play, Do. Technology, Innovation, and Organization* (Oxford: Oxford University Press).

Dosi, G. (1982) Technological paradigms and technological trajectories: a suggested interpretation of the determinants and directions of technical change, in: *Research Policy*, 11(3), pp. 147–162.

Edquist, C. (1997) *Systems of Innovation: Technologies, Institutions and Organizations* (London: Pinter).

Eurotechnology (Japan) (2005) *Japan's Game Industry*. Available at http://eurotechnology.com/store/jgames/index.html (accessed 20 March 2008).

Fuchs, G. & Shapira, P. (Eds) (2005) *Rethinking Regional Innovation and Change: Path Dependence or Regional Breakthrough* (London: Springer).

Garud, R. & Karnoe, P. (2001) *Path Dependence and Creation* (London: Lawrence Erlbaum Associates).

GEM (2005) Global Entrepreneurship Monitor 2005, Executive Report. Available at http://www.gemconsortium.org/download/1154207911515/GEM_2005_Report.pdf (accessed 13 February 2008).

Goto, A. (2000) Japan's national innovation system: current status and problems, *Oxford Review of Economic Policy*, 16(2), pp. 103–113.

Haddon, L., Mante, E., Sapio, B., Kommonen, K.-H., Fortunati, L. & Kant, A. (2005) *Creative User-Centered Design Practices: Lessons from Game Cultures* (London: Springer).

Harada, N. (2004) Productivity and entrepreneurial characteristics in new Japanese firms, *Small Business Economics*, 23(4), pp. 299–310.

Harada, N. (2005) Potential entrepreneurship in Japan, *Small Business Economics*, 25(3), pp. 293–304.

Hemmert, M. (2005) Japanese science and technology policy in transition: from catch-up to frontrunner orientation, in: C. Storz (Ed.) *Small Firms and Innovation Policy in Japan*, pp. 33–56 (London: Routledge Curzon Press/ Routledge Contemporary Japan Series).

Hitotsubashi Review (2005) Nihon no start ups (Start-ups in Japan), special issue (*tokushû*) 53(1, Summer) (in Japanese)

Hwang, W. S. (2002) *Nichibei Chûshô Kigyô no Hikaku Kenkyû* (Comparative research on SMEs in Japan and the USA) (Tokyo: Zeimu Keiri Kyôkai) (in Japanese).

Iwabuchi, K. (2002) 'Soft' nationalism and narcissism: Japanese popular culture goes global, *Asian Studies Review*, 26(4), pp. 447–469.

Jeppesen, L. B. & Molin, M. J. (2003) Consumers as co-developers: learning and innovation outside the firm, *Technology Analysis & Strategic Management*, 15(3), pp. 363–383.

JILPT (Japanese Institute for Labour Policy and Training) (2005) JILPT Kontentsu sangyô no koyô to jinzai ikusei (Contents industry and personnel development), *Rôdô Seisaku Kenkyû Hokokusho*, No. 25 (in Japanese).

Kawai, H. & Urata, S. (2002) Entry of small and medium enterprises and economic dynamism in Japan, *Small Business Economics*, 18(1–3), pp. 41–51.

Kishida, R. & Lynn, L. H. (2005) Restructuring the Japanese national biotechnology innovation system: prospects and pitfalls, in: C. Storz (Ed.) *Small Firms and Innovation Policy in Japan*, pp. 111–137 (London: Routledge Curzon Press/Routledge Contemporary Japan Series).

KKK (*Kokumin Seikatsu Kinyû Kôko*) (2006) 2005nendo shinki kaigyô jittai chôsa kekka (Results of the survey on start-ups of 2005), *Chôsa Geppô*, April, No 540, pp. 5–15 (in Japanese).

Kohashi, R. & Kagono, T. (1995) The exchange and development of images: a study of the Japanese video games industry, Kose University Discussion Paper Series 9527, Research Insitute for Economics and Business Administration.

Kuratani, M. (2005) Kigyôka seishin to iu konseputo, tokuni toshi no yakuwari to no sôgô sayô ni tsuite (The concept of entrepreneurship. Under special consideration of the role of cities and mutual compatibilities), *Chûshô Kigyô Sôgô Kenkyû*, August, pp. 18–36.

Lundvall, B.-A. (Ed.) (1992) *National Systems of Innovation* (London: Pinter).

Lundvall, B.-A., Johnson, B., Andersen, E. S. & Dalum, B. (2002) National systems of production, innovation and competence building, *Research Policy*, 31(2), pp. 213–231.

Maeda, N. (2001) Missing link of national entrepreneurial business model. In: *Management of Engineering and Technology, '01 Portland International Conference PICMET*, Vol. 2, pp. 85–99.

Malerbra, F. (2006) Innovation and the evolution of industries, *Journal of Evolutionary Economics*, 16(3), 3–23.

Malerba, F. & Orsenigo, L. (1996) The dynamics and evolution of industries, *Industrial and Corporate Change*, 5(1), pp. 51–87.

Marinova, D. & McAleer, M. (2003) Nanotechnology strength indicators, *Nanotechnology*, 14, pp. R1–R7.

McGuire, J. & Dow, S. (2003) The persistence and implications of Japanese *keiretsu* organisations, *Journal of International Business Studies*, 34(34), pp. 374–389.

METI (Japanese Ministry of Economy, Trade and Industry (2001) The twenty-eighth survey of industrial economic trends, White Paper, METI, Tokyo.

METI (2002) Trends in Japan's industrial R&D activities – principal indicators and survey. Available at http://www.meti.go.jp/policy/tech_research/indicator/english1411.pdf (accessed 21 February 2006).

METI (2006) Gêmu sangyô senryaku: gêmu sangyô no hatten to miraizô (Strategies in game software: development and perspectives of the game software sector). Available at http://www.meti.go.jp/press/20060824005/game-houkokusho-set.pdf (in Japanese) (accessed 1 June 2007).

Müller, C., Fujiwara, T. & Herstatt, C. (2004) Sources of bioentrepreneurship: the cases of Germany and Japan, *Journal of Small Business Management*, 42(1), pp. 93–101.

Nakagawa, K. (1999) Japanese entrepreneurship: can the silicon valley model be applied to Japan?, Available at http://iss-db.stanford.edu/pubs/10062/Nagakawa.pdf (accessed 21 February 2006).

NASDAQ (2002) The relentless pursuit of better ideas is fundamentally human – NASDAQ 2002 Annual Report. Available at http://www.nasdaq.com/investorrelations/ar2002/pdf/NDQ_AR_2002_complete.pdf (accessed 20 March 2006).

Nelson, R. R. (1993) *National Systems of Innovation: A Comparative Study* (Oxford: Oxford University Press).

Nelson, R. R. & Sampat, B. N. (2001) Making sense of institutions as a factor shaping economic performance, *Journal of Economic Behavior & Organization*, 44(1), pp. 31–54.

Nonaka, I. & Takeuchi, H. (1995) *The Knowledge creating Company – How Japanese Companies create the dynamics of Innovation* (New York: Oxford University Press).

OECD (Organization for Economic Cooperation and Development) (1998) *The Software Sector: A Statistical Profile for Selected OECD Countries* (Paris: OECD).

OECD (1999) *Managing National Innovation Systems* (Paris: OECD).

OECD (2006) *Policy Brief, July 2006* (Economic Survey of Japan) (Paris: OECD).

Okamuro, H. (2002) Recent changes in Japan's small business sector and subcontracting relationship, in: *Asian Productivity Organisation 2002, Strengthening of Supporting Industries*, pp. 40–53.

Pagulayan, R. J., Keeker, K., Wixon, D., Romero, R. L. & Fuller, T. (2003) User-centered design in games, in: J. A. Jacko & A. Sears (Eds) *The Human-Computer Interaction Handbook: Fundamentals, Evolving Technologies and Emerging Applications*, pp. 883–906 (Mahwah, NJ: Lawrence Erlbaum Associates).

Prügl, R. & Schreier, M. (2006) Learning from leading-edge customers at The Sims: opening up the innovation process using toolkits, *R&D Management*, 36(3), pp. 237–250.

Pyke, F. & Sengenberger, W. (1990) Introduction, in: F. Pyke, G. Becattini & W. Sengenberger (Eds) *Industrial Districts and Inter-firm Cooperation in Italy*, pp. 1–9, International Institute for Labour Studies, Geneva, 2005.

Ratliff, J. M. (2004) The persistence of national differences in a globalizing world: the Japanese struggle for competitiveness in advanced information technologies, *The Journal of Socio-Economics*, 33(1), pp. 71–88.

Schrage, M. (1999) *Serious Play. How the World's Best Companies Simulate to Innovate* (Cambridge, MA: Harvard Business School).

Schumacher, D., Legler, H. & Gehrke, B. (2003) Marktergebnisse bei forschungsintensiven waren und wissensintensiven dienstleistungen: außenhandel, produktion und beschäftigung: (market results of research-intensive goods and knowledge-intensive services: foreign trade, production and employment. Available at http://www.diw.de/deutsch/produkte/publikationen/materialien/docs/papers/diw_rn03-04-25.pdf (accessed 12 February 2006) (in German).

Shintaku, J., Tanakaz, T. & Yanagawa, N. (Eds) (2004) *Gêmu Sangyô no Keizai Bunseki* (Economic Analysis of the Game Software Sector) (Tokyo: *Toyo Keizai Shinposha*) (in Japanese).

Storz, C. (2005) Cognitive models and economic policy: the case of Japan, in: C. Storz (Ed.) *Small Firms and Innovation Policy in Japan*, pp. 82–108 (London: RoutledgeCurzon/Routledge Contemporary Japan Series).

Storz, C. (2006a) Pfadabhängigkeit und Pfadgenese – die japanische Spieleindustrie (path dependency and path genesis – the Japanese gaming industry), in: W. Pascha (Ed.) *Herausforderung Ostasien, ZfB Special Issue* (3) (Wiesbaden: Gabler Verlag) (in German), pp. 69–86.

Storz, C. (2006b) Japan's innovation system and 'cool' industries: what does the game software case teach us? (Research Report, Japan Institute of Labour Policy and Training), June. Available at http://www.jil.go.jp/english/reports/visiting.html (accessed 20 March 2008).

Storz, C. (2008) Dynamics in innovation systems: evidence from Japan's game software industry. Research Policy, forthcoming.

Storz, C. & Frick, S. (1999) Sponsored spin-offs in Japan – Anregungen für die deutsche mittelstandspolitik? (a stimulus for policies towards SMEs and family businesses in Germany?), *List-Forum*, 25(3), pp. 310–327 (in German).

Streeck, W. & Thelen, K. (2005) Introduction: institutional change in advanced political economies, in: W. Streeck & K. Thelen (Eds) *Beyond Continuity: Institutional Change in Advanced Political Economies*, pp. 1–40 (Oxford: Oxford University Press).

Sunagawa, K. (1997) Nihon gêmu sangyô ni miru kigyôsha katsudô no keiki to gijutsu senryaku (Development and technical strategies of Japanese entrepreneurs in the game software industry), *Keieishigaku* (Japan Business History Review) 32(3) pp. 1–27.

Sundbo, J. (2000) Organization and innovation strategy in services, in: M. Boden & I Miles (Eds) *Services and the Knowledgebased Economy*, pp. 109–128 (London: Continuum).

Thomke, S. H. (2003) *Experimentation Matters: Unlocking the Potential of New Technologies for Innovation* (Cambridge, MA: Harvard Business School Press).

von Hippel, E. (1986) Lead users: a source of novel product concepts, *Management Science*, 32(7), pp. 791–805.

West, J. & Gallagher, S. (2006) Key challenges of open innovation: lessons from open software, *R&D Management*, 36(3), June 2006, pp. 319–331. Available at http://www.joelwest.org/papers/westgallagher2004.pdf (accessed 20 March 2008).

Yanagawa, N. (2004) The innovation structure learning from the game industry, in: J. Shintaku, T. Tanaka & N. Yanagawa (Eds) *Economic Analysis of the Game Software Sector* (Tokyo: Toyo Keizai Shinposha). (Japanese: *Gemu sangyo ni manabu inobeshon kozo*, in: *Gêmu Sangyô no Keizai Bunseki*) (in Japanese).

Yukawa, K. (2003) A cluster of internet companies in Tokyo – review of Bit Valley (Tokyo: Fujitsu Research Center). Available at http://www.fri.fujitsu.com/en/erc/people/yukawa/no3.pdf (accessed 20 March 2008).

The Expected Roles of Business Angels in Seed/Early Stage University Spin-offs in Japan: Can Business Angels act as Saviours?

MASANOBU TSUKAGOSHI

Introduction

In recent years, university spin-offs (USOs) have been regarded by Japanese academia as a vital innovation engine. The commercialization of academic research activities has become an integral part of a once-distanced 'ivory tower' tradition, in both academic and financial terms. In the main, this shift has occurred because of recent university reforms, such as the privatization of national and public universities in Japan. According to the Ministry of Economy, Trade and Industry (METI), USOs are expected to be a means of realizing continuous and subversive innovations from intellectual seeds generated by Japanese universities. The Japanese government has promulgated various regulations, grants and infrastructure projects to support USOs as part of a general policy of fostering academic innovation and, consequently, of revitalizing the stagnant Japanese economy.

As these new ventures mature beyond the university laboratory and enter the real world of commercialization, the true test of their economic viability comes into view. One challenge faced by these new ventures emerging from an environment that has traditionally been insulated from the winds of commercial interests is to amass an operational team and attract the financing in order to grow. In other countries, most notably the United States, business angels have played a critical role at this stage – providing strategic advice, valuable networking introductions and early funding. In Japan, where the venture start-up community is relatively new, the role and effectiveness of angels in general – and in supporting USOs in particular – remains unclear.

This study explores more closely the environmental factors surrounding the USO community in Japan. It then outlines the characteristics and roles of business angels in the United States – arguably the most established of informal venture capital communities. The supply and demand of various financial *and* non-financial supports in Japan are examined, including the present activity of Japanese business angels. The existence of a gap between various non-financial needs of the USOs in Japan and the lack of non-financial value-added support commonly available to them is then analysed. Finally, there is a discussion of the possibility of Japanese business angels acting as a 'synergistic plug' designed to fill the gap between need and support and thus direct Japanese USOs more effectively to success.

Research Questions and Proposition

The overarching objective of this study is to clarify the potential roles that business angels can play in support of Japanese USOs in the Japanese venture capital (VC) community generally and in support of Japanese university spin-offs (USOs) in particular. Towards this objective, this essay addresses the following research questions:

- What is the current role of business angels in the Japanese VC community?
- What is the current role of business angels in support of Japanese USOs?
- How might this role be developed in future and to the benefit of both parties in seed/early stage USO ventures – above all, in respect of addressing the non-financial/management support needs of USOs?

Overall, this study explores and analyzes the proposition that, in Japan, a gap exists between the non-financial needs of USOs and the availability of such non-financial support to these USOs. It furthermore explores the potential of Japanese business angels to act as a 'synergistic plug' in order to fill or bridge this gap. Details of this analysis are presented below.

Context: The Current and Emerging Role of Business Angels in Japan

The purpose of this initial contextualizing discussion is to define some of the key terms relevant to our analysis and to outline some of the policy changes relevant to defining the role and scope of business angels in respect of Japanese USOs.

Defining USOs

University ventures, most narrowly defined, are incorporated entities formed to commercialize intellectual property rights, new technologies and/or business models that emerge from university research activities. In the context of this current study, a broader definition of a USO is adopted to include ventures that collaborated with university research initiatives, utilized university facilities (including incubation centres), or otherwise in-licensed technologies from a university (METI, 2006a).

Defining Business Angels

While identifying a USO is relatively clear, identifying a 'business angel' is not so straightforward. Some individuals may merely supply funding, while others may provide advice or networking introductions – for equity, for profit, or for free. In this current study the definition of 'business angels' is limited to that of value-adding individual investors who provide risk capital directly to new ventures in which they have no family connection, and contribute their commercial skills, entrepreneurial experience, business know-how and contacts through a variety of hands-on roles (Mason & Harrison, 1995). These informal venture capital (VC) investors play a vital role in supporting, both financially and non-financially, seed/early stage start-up developments. Nevertheless, and despite their relative importance, few sources of collective information on business angels are available in Japan owing (mainly) to their relative anonymity.

Compared to the role of business angels in the USA there has (hitherto) been little data on – and even less analysis of – the Japanese business angel community. However, their existence and potential impact was suggested in a 2002 survey by the Organization for Small and Medium Enterprises and Regional Innovation in Japan (SMRJ), whose report defines business angels as 'private individuals who provide entrepreneurs and early-stage businesses with both financial support, through capital investments, and various non-financial supports for business growth' (SMRJ, 2002a: 3). The SMRJ reached this definition after studying known definitions of business angels, including that offered by Harrison and Mason (1995; as cited above). Hence, combined reference to such sources should provide the analysis here with comparable data on investor characteristics and investment profiles in order to assess subsequently the potential future role of business angels in Japan.

Recent Policy Developments in Respect of USOs in Japan

The Japanese government has been working hard to support both USOs and promote the business angel community. The execution of the Science and Technology Basic Plan (First Phase) in fiscal year (FY) 1996 is often recognized as the beginning of policy developments for the recent university-industry collaboration in Japan. Between FY1996 and FY2000 this plan dedicated 17 trillion yen towards promoting university-industry collaborations, some of which resulted in USOs. METI and the Ministry of Education, Culture, Sports,

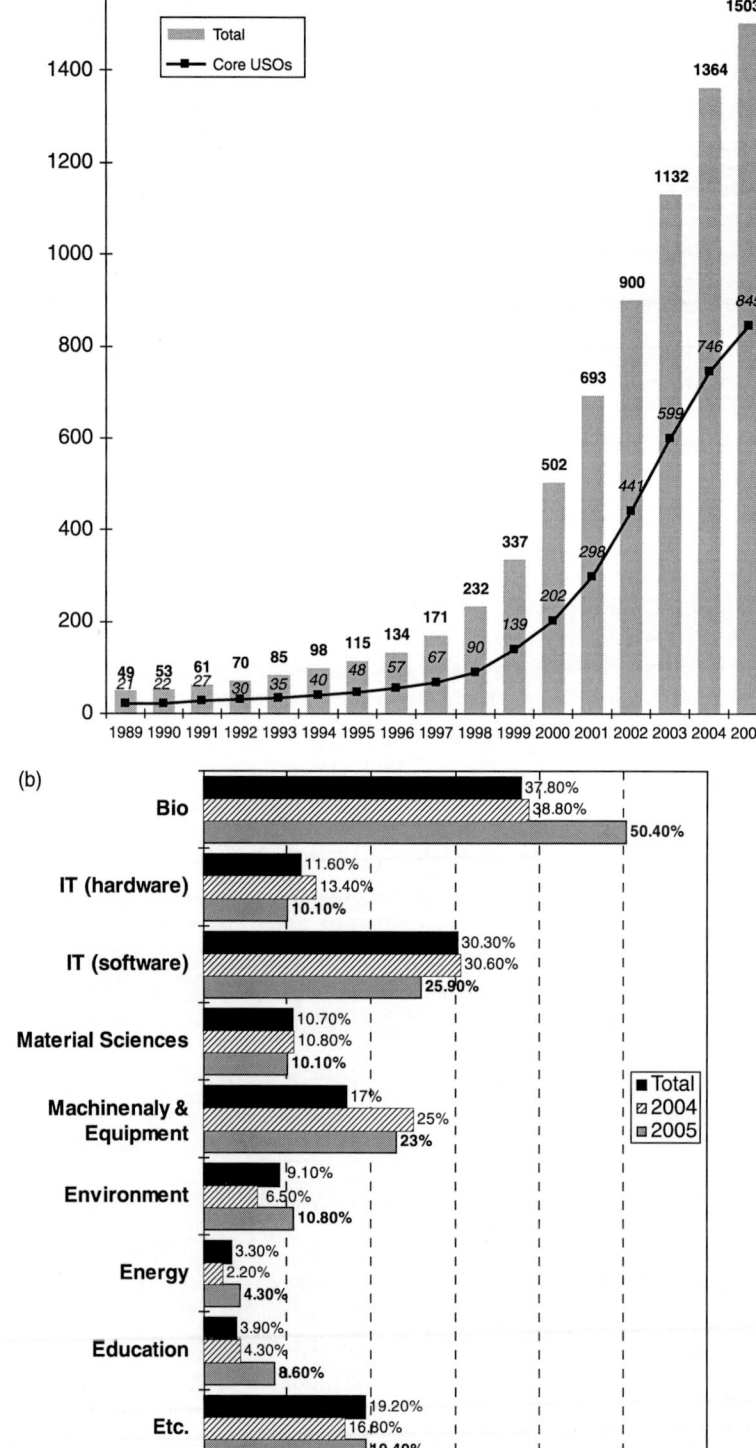

Figure 1. (a) Number of newly established USOs (fiscal year base). (b) Field of technology.

Science and Technology (MEXT) executed various regulatory efforts and grant programmes in support of USOs emerging from these efforts, such as a pre-venture grant programme under MEXT and technology licensing office (TLO) support programmes under SMRJ/METI.[1]

The second phase of the Science and Technology Basic Plan started in 2001 with a budget reaching 24 trillion yen through to FY2005. In the same year, the then economic minister Hiranuma introduced his target of establishing 1,000 USOs by FY2004, known as the '*Hiranuma* Plan'. To further promote industry-academic collaboration, along with revitalizing the stagnant Japanese economy, METI executed the Industrial Cluster Plan in FY2001, followed by the Intellectual Cluster Plan (produced by MEXT) in FY2002. The Japanese government released the Intellectual Property Basic Law in FY2002 to foster new measures and incentive programmes in intellectual property (IP) developments – a policy shift outlined by Taplin in this current collection. This change of law in respect of IP was followed by the privatization of national and public universities in FY2004, thus further accelerating the establishment of USOs in Japan.[2]

According to METI, USOs in Japan have totalled 1,503 companies in FY2005, of which core USOs have amounted to 845. Ever since establishing university technology licensing offices from 1998, the number of newly established USOs has dramatically increased at a rate of more than 200 start-ups per year (see Figure 1).

It is estimated that the total economic impact of Japanese USOs will reach 364.2 billion yen, providing 25,858 new jobs (METI, 2006a). The most popular field of technology among USOs is biotechnology (50.4 per cent), followed by information technology (IT)/software (25.9 per cent), machinery and equipment (23.0 per cent), IT/hardware (10.1 per cent) and material science (10.1 per cent). As suggested in Figure 1, non-technology related USOs in fields such as education and consulting accounted for 19.4 per cent of start-ups (METI, 2006a: 4–10).

The number of universities with USOs increased from 231 in FY2004 to 248 in FY2005, with a noticeable increase in liberal arts schools (METI, 2006a: 11–22). The top five universities in terms of cumulative number of USOs are the University of Tokyo (92 USOs), Waseda University (75 USOs), Osaka University (71 USOs), Kyoto University (59 USOs) and Tsukuba University (57 USOs). Owing to the greater number of existing universities in large prefectures, Tokyo, Osaka, Kanagawa, Kyoto and Fukuoka are the top five prefectures in terms of the cumulative number of USOs. In 2005, non-central regions such as Chubu, Chugoku, Shikoku and Kyushu indicated a noticeable increase in the number of USO establishments over the national average. As of March 2005, 16 USOs filed for an initial public offering (IPO), of which ten firms were biotech-related (METI, 2006a:24). According to a recent METI survey, 50.9 per cent were R&D and seed/early-stages ventures, and over 70 per cent of the start-ups were still running at a loss on an annual basis (METI, 2006a: 27–28).

Although the FY2006 survey results (compared to similar data from FY2005) have shown either progress or improvement in 33.8 per cent of the overall USO cases surveyed, around 66 per cent of these ventures reported no progress, or even a deterioration in their business operations. This suggests that there is room for improvement in expediting the commercialization process of their technologies (METI, 2006a: 29–30). In order to achieve a newly set target to make 100 USOs publicly owned by FY2010, the Japanese government has started examining required improvements and policy measures to meet this target.

The 'Angel Tax' Measure

To foster private investments in venture businesses, the Japanese government introduced a tax incentive measure in 1997. The so-called 'Angel Tax Treatment' allowed a three-year carry-forward period on capital losses generated by investing in pre-registered qualified venture businesses. After amending the measure a few times since introduction in order to fit the investment climate more precisely, the government has recently disclosed the expansion of tax measures to extend the carry-forward to March 2009, and widen the subject of qualified businesses from technology-oriented to business model-driven ventures.

The Emerging Significance of Business Angels

The type of government tax incentives for investment outlined here together with the increasing popularity of USOs suggests that a thriving business angel community would play a value-adding role in guiding Japanese USOs on the path to business success. However, what approach should the Japanese business angel community adopt? One option is to adapt from approaches established by business angels in the United States.

A Business Angels Model: Informal Venture Capital in the USA

The US business angel community is a well-defined informal VC market and represents the most established model for this field of investment activity. According to a recent survey taken by Center for Venture Research at the University of New Hampshire, the US angel investor market grew steadily in 2006 to US$25.6 billion, an increase of 10.6 per cent over 2005 and representing sustainable growth (Sohl, 2007). Takahashi (2000) has suggested that one out of four to five start-ups in the USA receive the support of business angels.

Indeed, it appears that an informal VC market has played a major role in filling the equity gap that exists for seed/early stage ventures (Harrison & Mason, 1992). A typical investment size of this equity gap ranges between 50,000 and 500,000 US dollars, the most popular segment for business angels (Yokota, 2004). Although this investment segment was shared by classic or hands-on venture capitalists – as it is still is, to a significant extent – the recent shift of investment focus to later stages by the overall VC community created such an equity gap. Over the past decade, VC funds in the USA have grown in both investment size and number of managing partners. As VC funds become more institutionalized, so their investment focus has shifted to later stage development in order to shorten their investment horizon and to capture gains early – a trend outlined by Ibata-Arens in this current collection. In fact, the average investment size per deal by VC funds in the USA increased from US$2.3 million in 1987 to US$4.7 million in 1997, and recently to US$7.5 million in 2006 (Oh, 2001; note 1).

According to Harrison *et al.* (1997), a typical US business angel is a male in his late 40s, who has postgraduate qualifications and has relevant business experience. Business angels not only often invest money but also supply experience-based skills and business know-how. It is common to see a business angel on the board of directors, and unofficially offering consulting services. On occasion, an active angel investor might work full-time or part-time for the companies he invests in (Harrison *et al.*, 1997). To paraphrase from Mason & Harrison (1995), the opportunity to be involved in an entrepreneurial venture is a significant motivating factor for US-based angels when deciding on informal venture capital investments, as well as a means of protecting their investment. Business angels seldom fund start-ups to exit. At some strategic point, they pass on the development task to venture capitalists (VCs). According to a survey conducted by George Washington University (cited in Akah & Stanco, 2005), 94 per cent of VCs responded that angels are beneficial to the venture industry.

Of course, not all informal investors are quality angels. Some unfortunately do not possess the required set of skills to support start-ups properly. These business angels may be transformed into vultures or devils – offering improper advice, distracting the core team, or taking over the investment venture. However, one should assume that most are inherently interested in nurturing the growth of their invested entrepreneurs and venture business. One can conclude that this active and informal venture capital market is an indispensable factor in the vital entrepreneurial economy in the US (Oh, 2000).

Analysis: Framework

Against the background of environmental factors outlined thus far in this study, there now follows an examination of the investment trends and styles in respect of Japanese institutional venture capital – as indicated above, one of the critical financing sources for USOs, followed by other financing sources such as governments and corporations. In this section, there is an analysis of whether non-financial value-added support properly exists in the context of seed/early-stage venture development. There is also a discussion of the current role of business angels in Japan in order to assess the effectiveness of their activities. The study then examines the existence of a gap between supply and demand of non-financial support and considers the potential of Japanese business angels to bridge this gap.

Analysis 1: Investment Stages and Styles of Venture Capitalists (VCs) in Japan as a Major Financing Source for USOs

Success in seed/early stage investments often requires more than just money. Hands-on VCs, also known as 'classic' VCs, provide their portfolio companies with various non-financial support, ranging from business planning to business development. These VCs often play an active role on the Board of Directors and so navigate their portfolio companies to contain downside risk while adding value to their investment. In the USA, hands-on VCs act both as a risk money provider and as an interest-sharing partner, who provide various types of managerial support. As Kirihata (2003) points out, of the managerial support provided by VCs the most important functions include utilizing personal networks and managing external relationships – such as supplying temporary help, expanding human networks and providing informative networking opportunities. In short, hands-on VCs seek to increase the corporate value of portfolio businesses through providing a comprehensive range of managerial support in their role as investors and/or partners.

Japanese VCs are generally viewed as passive investors, more comfortable with later stage investments. In large part, their activities are influenced by the source of limited partner funding. As of 1999, 54 per cent of the overall financing source for Japanese VCs was corporations, followed by banks at 18 per cent and insurance companies at 17 per cent (SMRJ, 2002b: 23–26). During the same period, the most popular investment stage for Japanese VCs was expansion at 50 per cent, followed by mezzanine (between expansion and buy-outs/IPO) at 19 per cent, with start-ups comprising only 15 per cent. Thus, Japanese VCs have appeared to prefer the later stages of investment.

In recent years, however, Japanese VCs have started looking more to earlier stage investing. There is anecdotal evidence, to support the view that a growing number of Japanese firms are focusing (and perhaps more aggressively) at new venture start-ups, including USOs. A recent survey by the Venture Enterprise Center (VEC) supports this view. During FY2005, the 105 VC respondents surveyed invested 196.8 billion yen into 2,759 companies, of which 48.7 per cent had been incorporated less than five years prior to the investment (VEC, 2006: 2–3). Although the survey does not capture the entire VC industry in Japan, the figures cited are believed to be illustrative of an overall trend.

Japanese VCs have now started forming new funds with a focus – or at least a scope – that includes USOs as a target. As of FY2006, 55 venture funds totalling 84.9 billion yen received matching funds under the METI investment programme for fostering new business exploitation. Of these, nine funds totalling 17.5 billion yen specifically target the USO sector. Although no aggregate data is available, there are other new VC funds targeting USOs that have not sought matching funds from the central government, some of which focus on regional opportunities. According to METI (2006b), Japanese VCs were ranked as the top source for financing current USOs, investing 397 million yen per company on average at the seed/early R&D stage – a level of investment followed by private corporations with 283 million yen and government grants of 96 million yen. Although this analysis alone cannot confirm ample funds to fill in the entire equity gap in the seed/early investment stage, Japanese VCs appear increasingly willing to fund USOs at an early stage.

The Current Profile of Japanese VCs

Against the background of these apparent trends describing investment interest among established VCs, is there still a role for the Japanese business angel? As will be argued subsequently, just because VCs have started to focus more on early stage investments does not mean that new USOs can expect to receive high-risk funding as well as critical business advice from these investors. Indeed, I argue from experience that the business angel remains an important element of the venture even after the professional investor places a bet.

Unlike in the USA where private VCs dominate, the majority of Japanese VCs are subsidiaries of either financial institutions or large corporations: as outlined in our discussion above, the investment style of Japanese VCs differs markedly from that of their US counterparts. Individual venture capitalists working for Japanese VC firms are often referred to as *salaryman* venture capitalists, since many of them are either visiting or temporarily transferred personnel from a parent company. Drawing on VEC (2006) data, it appears that although 32 per cent of the individual capitalists in financial institution-related VCs and 67 per cent of those in corporate VCs claim over 20 years of previous professional experience, their VC experience is relatively limited, in fact. Of those VCs managing venture money, 62 per cent have less than five years experience in the VC space while 20 per cent have between five and ten years' experience. The most common function while working at a VC firm/subsidiary was deal analysis and execution at (87 per cent), followed by due diligence and planning at 83 per cent. Contradicting a common view that Japanese VCs often do not provide hands-on support, 72 per cent of those surveyed claimed to have experience in managerial support (VEC, 2006: 71–74).

Turning our eyes to actual investments by Japanese VCs, nearly 90 per cent of those surveyed stated that they provide managerial support to their portfolio companies. However, only 58 per cent are dispatched as directors while 53 per cent provide financial and administrative support. This is surprisingly low in light of the increasing levels of investment in early stage companies, and the fact that many VCs themselves are related to financial institutions. It is significant that nearly 80 per cent of the VCs who recognized the need of managerial support could not provide such support due to a lack of proper human resources (VEC, 2006: 77–78). One possible explanation for this shortfall in support lies in the relatively short length of individual service in a corporate VC structure, for acquiring relevant managerial capability often requires years of professional VC experience and hands-on know-how.

To summarize this first section of analysis, the above data reveals that the understanding of 'hands-on support' among venture capitalists in Japan is limited by their rather passive investment culture. This suggests that their seed/early stage investments are likely to receive only limited non-financial support. Although there are signs of quality hands-on VCs in Japan, their availability and number relative to the overall VC industry is limited (Jinza, 2005). Most VCs in Japan adopt a less individual approach and more organizational approach to investment, which perhaps is in agreement with the established investment style of the Japanese VC community. It can be argued that this organizational approach to investment might be of limited value to seed/early stage investments.

Analysis 2: Government Grants and Private Corporations as Sources of R&D Funding and Knowledge Supporting USO Development

The Japanese government is intent on increasing the number of new USOs. This objective has become a critical element of university reforms, as outlined subsequently by Debroux in this current collection. To illustrate: METI alone has consistently built a sizeable budget specifically related to USOs over the past few years – 43.4 billion yen in FY2003, 45.1 billion yen in FY2004, 50.9 billion yen in FY2005, and 52.8 billion yen in FY2006. These budgets have been applied to various programmes from R&D grants to human resources (HR) development (METI, 2003, METI, 2005). In addition, MEXT budgeted 27.5 billion yen in FY2004 and 29.4 billion yen in FY2005 for university-industry collaboration activities, some of which were USO-related. The METI figures alone suggest that an average sum of approximately 34 million yen was available for each of 1,503 USOs counted in FY2005. Though generalized here, this funding availability is significant given that more than half of these USOs are still at their seed/early R&D stages.

Furthermore, a comparative study of entrepreneurship in Japan and the USA's Silicon Valley also suggested that Japanese venture businesses enjoyed better access to diverse funding sources, particularly bank loans and government financing (Suzuki, *et al.*, 2002). As there are other public financing sources such as regional governments are also available, Japanese USOs seem to have various means to obtain financing beyond VCs and business angels.

In addition, the recent and protracted economic stagnation in Japan has forced many large corporations to streamline their internal R&D management. As a consequence, large corporations have been outsourcing some R&D processes to university laboratories and consistently building up collaboratively funded research with universities. To illustrate: the MEXT homepage tells us that Japanese universities received approximately 21.6 billion yen in FY2003 and 26.4 billion yen in FY2004 from private corporations.[3] Although this corporate funding primarily targets collaborative R&D, such activities sometimes result in establishing USOs, as the study by Debroux (in this current collection) illustrates. Furthermore, these R&D collaborations can act as a wellspring of experiences and knowledge in the university laboratories for future technology commercialization and thus sustain an on-the-job habitat for future technology entrepreneurs.

Analysis 3: Characteristics and Activities of Japanese Business Angels

There is a popular view that the business angel community is less developed in Japan than in other developed nations. Harada & Kutsuna (2002) reasoned that in Japan:

1) there is a relatively smaller percentage of business professionals wealthy enough to become business angels;
2) few people succeed as entrepreneurs;
3) there is a less-than-functional market in respect of providing a reliable and efficient risk-and-return structure from seed to IPO;
4) there are insufficient tax incentives for business angel activities; and
5) the matching system between entrepreneurs and business angels is less developed.

The Current Profile of Japanese Business Angels

Although the business angel community as a whole might be relatively underdeveloped, SMRJ recently reported a potential of Japanese business angels as a vital supporter in the seed/early stage venture development. This survey (SMRJ, 2002a) covered both business angels (73 respondents) and venture businesses (160 respondents) and revealed that the typical profile of a Japanese business angel is male (85 per cent), 56 years old, and holding a bachelor's degree or better (70 per cent), with 10 per cent holding advanced degrees. The most popular fields of expertise of Japanese business angels are management and economics, followed by science and engineering – a ranking that is opposite that of US business angels (see discussion above). The historically lower pay scale and enduring lifetime employment system of Japanese corporate engineers might explain this contrast.

The SMRJ (2002a) report further states that the start-up experience of Japanese business angels was noticeably lower than that of their foreign counterparts: 51 per cent of those surveyed were experienced entrepreneurs, of which 30 per cent had experienced an IPO. To compare: 83 per cent of US business angels and 75 per cent of Canadian business angels were experienced entrepreneurs. About 60 per cent and 15 per cent of Japanese business angels still hold president/chief executive officer (CEO) and chairman positions, respectively. 67 per cent of these business angels still work in a corporation, with 8 per cent of them being retired. Although the the dominant male figure is similar in Japan and in the USA, becoming a business angel in Japan appears to require more time – possibly to accumulate sufficient wealth and experience. This supposition is supported by the older average age (about 10 years) of Japanese angels in comparison with their US counterparts.

Although some differences can be seen in the general profiles of business angels in Japan and in the USA, their economic figures are comparable. Japanese business angels earn on average 18.6 million yen per annum and have an average net worth of 390 million yen. Although these figures are higher than those of US angels, factoring higher income tax in Japan and real estate inclusion in the Japanese net worth would make the average financial position of Japanese business angels comparable to that of US angels (SMRJ, 2002a). The SMRJ report also points out how Japanese business angels invest on average about 13.4 per cent of their net worth, which is lower than the average figures for angels in other developed nations. The report further notes that the inclusion of real estate in net worth might be a reason for this difference.

In terms of geographical distribution, the majority of Japanese angels in the survey are located in the metropolitan Tokyo area (40 per cent), followed by Osaka Prefecture (8.3 per cent) and Kanagawa Prefecture (6.3 per cent). Although it may be premature to conclude firmly, the SMRJ survey suggests a potential nation-wide availability of business angels in Japan given that one-third of respondents live in non-prime regions. This is significant – and encouraging, since many recent USOs have been established in non-central regions.

In terms of investment decision making, Japanese business angels rated (at 52 per cent) the personal integrity, skills and capability of the entrepreneur as the highest investment factors, followed by the business growth potential at 50 per cent. In terms of motivating factors, 53 per cent of Japanese business angels echo their US counterparts by citing 'wanting to enjoy the growth process of companies' as the prime reason to decide on actual investment. Also noted were the expected investment return at 44.4 per cent and the social contribution level at 33.3 per cent. This trend fits potentially well to support well-funded USOs in Japan since business angels may have lesser opportunities to actually invest in the USOs.

Unsurprisingly, perhaps, 37.3 per cent of Japanese business angels becoming actively involved with their invested companies reported that wanting to enjoy the growth process was their main reason to get involved, followed by expecting investment returns (at 28.4 per cent). These active business angels seem to provide various non-financial support to the invested start-ups, such as:

- Advising on management strategy (ranks the highest, at 19.5 per cent).
- Monitoring management activity (14.3 per cent).
- Encouraging entrepreneurial activity (13 per cent).
- Advising on marketing (10.4 per cent).

The top position that these business angels actually held was as unofficial adviser (at 32.7 per cent). A directorship position was ranked third (at 21 per cent). Although different degrees of relationship become apparent from the SMRJ survey, around 30 per cent and 50 per cent respectively expressed their proactive interests in future investment opportunities.

In apparent contrast, business angels in other nations such as the USA and the UK have a mixture of financial and non-financial motives, with the prospect of financial gain as the single most important factor (Tashiro, 1999). Tashiro, using a separate survey, confirmed our assessment that financial gain is a minor reason for investment by Japanese business angels. As the informal venture capital market in Japan builds experience and success stories, Japanese business angels may become more financially oriented and thus more in balance with their non-financial motives. In general, however, it can be concluded that business angels in Japan as elsewhere share similar characteristics and interests in nurturing the development of their start-ups.

Analysis 4: The Needs and Challenges of USOs in Japan

There are many challenges in the different stages of a venture's growth. Metaphors such as 'The valley of death' or 'Darwinian Sea' are often used to depict such challenges. Although these challenges exist, enactment of the TLO Act, regulatory relaxation of employing professors and researchers, developments of start-up incubation facilities, privatization of national and public universities, along with increasing awareness within academic personnel, have all contributed to an increase in the number of USOs established in recent years (METI, 2006b:11). It appears that approximately 33 per cent of the USOs surveyed had not exercised, or appeared not to know about, market research. This suggests a tendency towards technology driven approaches to business development, although 52.9 per cent of the USOs surveyed have an IPO as their exit strategy (METI, 2006a: 31–32). As for funding in the seed/early-stage development, the most popular approach was to utilize their initial capital, funded internally and/or externally, and national/regional-government grants. Utilizing initial capital to weather the storm is not unusual for start-ups and is, in fact, often encouraged. Accessing government sources is also unsurprising since (as suggested earlier) ample public financing options are available for Japanese USOs. Given that more than half of the USOs in FY2005 were still at the seed/early R&D stages, the study will now look at various managerial challenges that these USOs are facing.

The Profile of Seed/Early-Stage USOs: HR Issues

Although recently relaxed, the legal process to establish a company is still a time-consuming administrative task in Japan. However, it is nothing compared to developing and growing the company thereafter. One METI report (2006b) highlighted three major challenges that current USOs faced: securing and nurturing human capital, fund raising, and finding new markets. In this particular report, the single most obstructive factor was the HR issue from inception to current stages. As most USOs in Japan are technology/research-driven, it was not surprising to note that 25% of their current CEOs were professors (14.8 per cent) and students/researchers (9.4 per cent), especially at seed/early stages. But this management structure is not without its concerns, as these CEOs have little or no business experience. 19.1 per cent of USOs have changed their CEO at least once since inception, of which 59.1 per cent reasoned that their development stage had changed. However, this admission does not necessarily mean that those USOs actually have found business savvy managers. In fact, only 9.9 per cent responded that they had found more appropriate individuals from the private sector, which indicates continuing difficulty in attracting quality executives.

Many USOs also seek individuals with a higher academic degree to oversee both R&D functions and business development. The most expected function for such individuals was business planning at 53.3 per cent, followed by new market development and fund raising at 44.3 per cent. The Chief Technology Officer (CTO) function was ranked the lowest at 18.9 per cent. These results can be explained perhaps by reference to 1) the technical nature of core technology requires such understanding even for business managers, and 2) the current CEOs being the likely inventors of the core technology and already acting CTOs. It can also be inferred from this result that the professor/researcher CEOs are in need of smart 'all-rounders' to fill in for labour-intensive R&D processes and business functions requiring relatively less experience, given the general lack of resources to hire personnel.

Also noted was the success rate to secure personnel. The rate for obtaining business managers for new market development was the lowest at 50.5 per cent, followed by management executives at 59.2 per cent. Unsurprisingly, perhaps, attracting R&D personnel was the most successful at 70.7 per cent, probably owing to their proximity to academic researchers and students. Their associated university and/or friends were the most popular route to seek human resources, ranking at around 70 per cent. However, it can be argued that business savvy managers and managerial executives are to be found elsewhere. Correspondingly, beyond the expected close networks, using VCs as a referral route ranks at only 1.9 per cent to 9.0 per cent and, with a success rate of 33.3 per cent to 66.7 per cent, the conclusion might be which that VCs are not necessarily an assured source of human capital.

Linking HR to VC Funding

However, and as mentioned earlier, the VC community remains the primary funding source for seed and early stage USOs, followed by private corporations and government grants. However, the most popular funding source seems to lack an appropriate level of non-financial support, as evinced in the HR difficulties outlined above. The METI survey in particular revealed the dissatisfaction that many funded USOs had with their VC support. Although many were still at early R&D stages, 22.8 per cent of USOs only received financing from VCs. The remainder received some sort of non-financial support, but of a relatively passive nature, such as advice on business planning. It is inevitable that, as a profit entity, VCs need to balance their non-financial involvement (a time factor) relative to actual investment size. Consequently, a small early-stage USO cannot expect much attention relative to much larger investments. However, the analysis developed here raises concerns about the success probability of those USOs invested in, especially during their early stages when such support is most critical. Ultimately, VCs themselves will need to cover the possible investment loss if their invested USOs fail, unless some improvement measures – internally and/or externally – are taken.

Implications

Although financial support is naturally on the wish list of most USOs, the analysis here has revealed that financial support seems readily available. Often this is actually provided to Japanese USOs because of the early stage investment interests of Japanese VCs in combination with various government policy measures to support USO development. Financing is important in both seed/early stages and in later development stages. However, non-financial support can be just as important as monetary support, and especially when the venture business is at its infant stage. It has been seen here that non-financial support is often lacking for seed/early stage USOs in Japan, as opposed to the relatively ample financing options. Consequently, it is possible to identify a distortion or 'gap' between the supply and demand equation of non-financial support to USOs.

In terms of profile it has also been found that business angels in Japan, though still underdeveloped as a community, have similar personal traits and interests in nurturing start-ups as their counterparts in other well-developed informal VC markets such as the USA. Surveys suggests that proactive Japanese business angels already exercise their potential in supporting venture start-ups in both financial and non-financial terms. It is argued here, therefore, that the professional infrastructure required for guiding seed/early stage

Japanese USOs to success can be improved with more proactive involvement by Japanese business angels.

In recent years, Japanese universities have built many venture incubation centres and USO support programmes. These have been portrayed as a valuable tool with which to transform Japanese academic society to a more dynamic and entrepreneurial one. However, these facilities and programmes are only as strong as their management components (Wolfe *et al.*, 2001). The most critical factor to the success of start-ups utilizing such programmes is to have personnel who can guide business decisions, provide experienced advice and develop the strengths of entrepreneurs (Drucker, 1985). I argue that quality business angels are capable to perform this function. However, and despite this possibility, effective collaboration between incubators and angels in the Japanese USO ecosystem appears to have been rare. I believe that an invisible hurdle between universities and business angels exists, described by reference to the following challenges at both ends of the spectrum.

The old bureaucratic-like management behaviour of universities – to do, or try to do, everything in-house – is looked upon as one of the causes of current USO ineffectiveness. Expressing old management styles continues in many academic institutions, probably because the government funds various costly propositions such as building a venture laboratory. As a consequence, Japanese universities are less pressured to manage such facilities with a business mindset, and furthermore, are prone to creating a self-contained, ineffective USO support infrastructure. As a result, fewer effective personnel might be hired or positioned to support USOs. Certainly, functional outsourcing is a low priority in such old-fashioned systems. Along with existed protectionism generated by vested rights and academic pride, the traditional university-style management might have created an invisible barrier to outside help.

Business angels are also required to be more visible and accessible to university management. In a recent interview, Mr Keisuke Yawata, Chairman of IAI Japan observed that Japanese business angels are less organizational and less motivated to disclose his/her existence.[4] A well-known Japanese business angel himself, Mr Yawata further pointed out that Japanese business angels are arguably less motivated by financial interests, as in other countries such as the USA – possibly because of cultural factors that tend to downplay one's status and wealth. Although often anonymous and geographically dispersed, business angels do exist nationally in Japan and are looking for new opportunities.

Matching universities and business angels in Japan therefore seems to require a separate 'lubricant' agent. One possibility is an 'organizational' relationship between these two parties. As business angels are inherently individualistic, they have less success in opening the door of the traditionally aloof academic world. To compensate for such drawbacks, the aforementioned IAI Japan has been gradually building a possible organizational alliance with some universities in the Kanto region. As trust building is the single most critical factor in angel activities, and remains so important in traditional Japanese society, utilizing the collective individual qualifications of angel members may become a powerful tool to bridge the gap between the two parties.

A more compelling option for an angel network or organization, perhaps, is to establish a business alliance with those institutional VCs already investing in USOs, especially the VCs dealing with problematic USOs. The vested interests of such alliances are much clearer than the earlier option since those VCs want to maximize investment returns, and angels seek to experience the nurturing process of USOs. The support of promising USOs by Japanese VCs remains essential to the growth of the USO community as a whole. As most USOs in Japan are technology-related and require heterogeneous personnel skills, Japanese VCs may ultimately perceive such investments as too risky, and will – unless some improvement measures are taken – turn to later stage investment opportunities. Although such argument is still premature, it is possible that the shifting of investment focus to later stages may occur if the VC industry dynamics in the USA can be a guide to what may happen in Japan.

In their role as potential saviours to USOs in Japan, there should be more attention given to the potential role of business angels generally, and particularly as part of an organizational approach to building strategic alliances among early stage supporters, as remains the norm in traditional Japanese society.

Conclusions

This study is by no means intended to criticize the Japanese VC community in their current association with USOs, or the Japanese government and universities in their efforts to develop policies and measures of university-industry collaboration. Rather, this paper suggests that both parties have the potential to play a critical role in fostering a shift in the focus of the Japanese economy from the large corporate to the more entrepreneurial. As mentioned above in this study, there are encouraging signs of quality hands-on VCs supporting start-ups in Japan. However, these remain the exception rather than the rule.

Japanese policy makers have been proactively supporting innovation, and promising results can be seen in some universities and research institutions across Japan. The factors for a successful venture are varied: quality of entrepreneurs, sound intellectual property portfolio development, the general economic environment, and a little bit of luck. All these factors play important roles in the business success of USOs. It is inevitable that, in the context of human interaction and new forms of cooperation, a degree of continuous trial-and-error must be allowed, enabling parties to find the most appropriate infrastructure and programmes for the environment in question. As a consequence, further research should look into cultural factors influencing business angel activities and their perceptions in relation to (relatively speaking) insular academic entrepreneurs in Japan.

As an early-stage venture capitalist and a humble angel practitioner myself, I will conclude this study by paraphrasing from the enduringly wise words of Peter Drucker (1985) who, when discussing 'entrepreneurial management' reminded us that, in standard institutional contexts, the 'entrepreneurial' element of this process is the key to success. However, he reminded us also that, in the venture business, it is 'management' that is key.

Acknowledgements

The author would like to take this opportunity to thank Dr Philippe Debroux, a Professor at Soka University, in providing him with this opportunity to extend his research efforts. He also would like to extend his appreciation to Mr Karl Ruping, his old friend and business colleague at incTANK Ventures, and to Mr. Sugiyama, a quality researcher at the University of Tokyo, who inspired and encouraged him in completing this research.

Notes

1 See the website for NVCA (National Venture Capital Association), available at http://www.nvca.org/ffax.html
2 See *Nikkei Business* (14 November 2005); Cabinet Office, Government of Japan, Office of Intellectual Property Strategy (2006).
3 See the website for the Ministry of Education, Culture, Sports, Science and Technology (MEXT). Available at http://www.mext.go.jp/b_menu/houdou/17/06/05062201.htm (in Japanese).
4 Interview with Mr Yawata is available at http://www.iai-j.com – a not-for-profit organization affiliated with International AngelInvestors in the USA).

References

Akah, U. & Stanco, T. (2005) Survey: the relationship between angels and venture capitalists in the venture industry, unpublished manuscript. Available at http://www.papers.ssrn.com/sol3/Delivery.cfm/SSRN_ID1093006_code339992.pdf?abstractid=1093006&mirid=1 (accessed 20 March 2008).

Cabinet Office, Government of Japan, Office of Intellectual Property Strategy (2006) Chiteki Zaisan Suishin Keikaku 2006 (Intellectual Property Promotion Plan 2006) (in Japanese).

Drucker, P. (1985) *Innovation and Entrepreneurship – Practice and Principles* (London: Heinemann).

Harada, N. & Kutsuna, K. (2002) Senzaiteki bijinesu enjeru to siteno chuushou kigyou keieisha (Small and medium enterprise managers as a potential business angels), Research Institute of National Life Finance Corporation Japan, Chosa Kiho (Seasonal Research Report), No. 63, p. 2 (in Japanese).

Harrison, R. & Mason, C. (1992) International perspectives on the supply of informal venture capital, *Journal of Business Venturing*, 7, pp. 459–475.

Harrison, R., Mason, C. & Nishizawa, A. (1997) *Bijinesu Enjeru no Jidai* (Original Title: Informal Venture Capital – Evaluating the Impact of Business Introduction Services) (Tokyo: Toyo Keizai Shinposha) (in Japanese).

Jinza, Y. (2005) *Nihon no Bencha Kyapitaru* (Japanese Venture Capital) (Tokyo: First Press) (in Japanese).

Kirihata, T. (2003) Fostering university-launched ventures and venture capitals – required value-adding capabilities of venture capitalists, *Journal of Mitsubishi Research Institute*, (42), p. 67.

Mason, C. & Harrison, R. (1995) Closing the regional equity capital gap: the role of informal venture capital, *Small Business Economics*, 7, pp. 153–172.

METI (2003) *Heisei 16 Nendo Sangakukan Renkei Kanren Yosan Ichiran* (FY2004 Budget Table Related to University-Industry Collaboration) (Tokyo: Ministry of Economy, Trade and Industry) (in Japanese).

METI (2005) *Heisei 18 Nendo Sangakukan Renkei Kanren Yosan Ichiran* (FY2006 Budget Table Related to University-Industry Collaboration) (Tokyo: Ministry of Economy, Trade and Industry) (in Japanese).

METI (2006a) *Daigaku-hatsu Bencha ni Kansuru Kiso Chousa*, (Report on Basic Research of University Spin-offs) (Tokyo: Ministry of Economy, Trade and Industry) (in Japanese).

METI (2006b) *Daigaku-hatsu Bencha no Seichou Shien ni Kansuru Chousa Houkokusho* (Research Report on Growth Support for USOs) (Tokyo: Ministry of Economy, Trade and Industry) (in Japanese).

Oh, S. (2000) Beikoku kigyouka keizai ni okeru bijinesu enjeru no yakuwari (Role of business angels in entrepreneurial economy in the United States), Sapporo University, *Keizai to Keiei*, 31(3), p. 49 (in Japanese).

Oh, S. (2001) Beikoku bijinesu enjeru no koudou youshiki – bencha kyapitarisuto tono hikaku (Activity style of US business angels – comparison to venture capitalist), Sapporo University, *Keizai to Keiei*, 32(2), p. 120 (in Japanese).

SMRJ (2002a) *Bijinesu Enjeru no Jittai Chousa Houkokusho* (Fact-finding Report on Business Angels) (Tokyo: Organisation for Small and Medium Enterprises and Regional Innovation) (in Japanese).

SMRJ (2002b) *Bencha Kigyou ni kansuru Kokunaigai no Chokusetsu Kinyuu (Toushi) Kankyou Joukyou Chousa Houkokusho* (Report on Domestic and Foreign Direct and Indirect Financing – Investment Environments for Venture Businesses) (Tokyo: Organisation for Small and Medium Enterprises and Regional Innovation) (in Japanese).

Sohl, J. (2007) The angel investor market in 2006: the angel market continues steady growth, Center for Venture Research, University of New Hampshire, *Media Relations News*, p. 1.

Suzuki, K., Kim, S.-H. & Bae, Z.-T. (2002) Entrepreneurship in Japan and Silicon Valley: a comparative study, *Technovation*, 22, pp. 595–606.

Takahashi, T. (2000) Waga Kuni ni okeru Oubeigata Bijinesu Enjeru no Kanousei – Kojin Tousika ga Seichou Kigyou ni hatasu Yakuwari (Potential of western-style business angels in Japan – role of individual investors in growing companies), Research Institute of National Life Finance Corporation Japan, *Geppo* (Monthly Report), pp. 10–11 (in Japanese).

Tashiro, Y. (1999) Business angels in Japan, *Venture Capital*, 1(3), p. 271.

VEC (2006) *Seisei 17 Nendo Bencha Kyapitaru-tou Toushi Doukou Chousa* (FY2005 Venture Capital Investment Trend Survey) (Tokyo: Incorporated Foundation Venture Enterprise Center)) (in Japanese).

Wolfe, C., Adkins, D. & Sherman, H. (2001) *Best Practices in Action – Guidelines for Implementing First-Class Business Incubation Programs* (Athens, OH: National Business Incubation Association).

Yokota, A. (2004) Kansai ni okeru bijinesu enjeru no Kasseika Saku – sougyouki no bencha kigyou shien wo mezashite (Activation measure of business angels in Kansai Region – aiming at supporting early-stage venture businesses), *Japan Research Review*, 3, pp. 68–69 (in Japanese).

Innovation in Japan: What Role for University Spin-offs?

PHILIPPE DEBROUX

Introduction

By traditional measures such as spending on research and the number of patents registered, Japan is doing quite well in comparison to the United States. However, this is not the case for the less tangible but more important measures such as the number of entrepreneurial start-ups, the level of venture capital funding, and the payback from patents. At the other end of the spectrum, the national innovation systems of India and China may not yet be capable of radical or breakthrough innovations. Nevertheless, they are likely to enter the market and offer an increasingly large range of innovative products and services. Those products may (initially) be regarded as inferior to Japanese ones; however, they have a level of technology sufficient to be launched successfully on world markets. Because they have a large domestic market composed of people who have just begun consuming, Indian and Chinese companies can build up experience domestically while improving the quality of their products, doing the same as Japanese companies did 30 years ago and South Korean companies thereafter. The open technology environment and the modularization trends prevalent today allow them to be competitive in many industries without having the managerial and technological expertise of western and Japanese companies. As a consequence, Japanese companies can find themselves squeezed between US and European companies and those from emerging Asian economies.

Adopting this perspective in Japan suggests that large companies are required to make significant efforts while a more entrepreneurial business environment must be created in order to boost small and medium sized enterprises (SMEs). However, many in the international community surely doubt whether Japan can develop such a vibrant entrepreneurial sector, citing Japan's group-oriented, consensus-driven society and an educational system that has never encouraged the attributes associated with entrepreneurship such as critical thinking, a positive response to change, initiative, and profit-orientation. The 'Silicon Valley model' – centred on constantly renewed dynamic interrelationships between large companies, public authorities, universities and venture capitalists – seems far away from the reality on the ground in Japan. Creating a new company may yet become a socially and professionally recognized career path for the graduates of the elite academic institutions in Japan. However, reforms have substantially improved the environment for entrepreneurial start-ups since the late 1990s, as explained already in several studies presented in this current collection. This may lead to the creation of a larger group of individuals ready to take higher risks and able to create better companies, thanks to a wider portfolio of technical and managerial expertise and better business connections.

Regional Variations

True, current economic recovery is skewed towards large cities, especially Tokyo, Nagoya and Osaka, whilst some regions fall behind. Many regional cities are close to technical bankruptcy and are bound to become a heavy financial burden for central government in the years to come. A vicious circle is forming again, with young Japanese people and companies once again attracted to the large cities, depriving the regions of the very human resources and sources of dynamism they need for their revival. Nevertheless, and despite current difficulties, there is a hidden dynamism in private universities and in the regions. There are public incubators in virtually all prefectures, as in many regional universities. They provide more affordable office space, telecommunications access and miscellaneous business services. Therefore, with the right incentives a substantial number of new ventures could be created that contribute to the renewal of local economies.

The Emerging Role of USOs

Although Japanese university spin-offs (USOs) are unlikely to play an important economic role in the short and medium term, their emergence is symptomatic of the changes in Japanese society and economy. Japanese authorities push strongly for the creation of elite start-ups in the hope that they would revive the national innovation system and create not only high-level jobs in themselves but, through a spillover effect, allow the creation of new jobs in the regions suffering from low growth and high unemployment. Japanese authorities have no intention (and anyway would be unable to put it into force, should they wish) to revive the kind of direct interventionist industrial policy adopted until the 1970s. What they seem to have in mind is the development of the kind of institutional setting with the right types of incentives based on market principles, as has been created in the United States since the Reagan era. In the mind of the public authorities this should allow Japanese companies, universities and research centres to emulate what the US did for instance in the semiconductor industry (exemplified by the SEMATECH semiconductor industrial project) or in biotechnology – themes developed respectively by Okada and by Ibata-Arens in this current collection.

Meanwhile, central government support is offered to existing industrial clusters and to the development of new ones all over the country (Tsukamoto, 2005). In these clusters, both large and small companies have an important role to play in order to develop conjointly a dynamic innovation system. The direct presence of the state is thus still strong in a number of them (Ibata-Arens, 2005), but this can be largely explained by historical reasons. Market principles should rule in principle with the minimum of bureaucratic intervention.

To illustrate: financial institutions are encouraged to start again to finance small companies, while miscellaneous financial and non-financial incentives programmes are put in place to encourage the creation of start-ups (METI, 2006). In order to have companies depart from indirect financing, the development of regional Over The Counter (OTC) markets is promoted in order to facilitate their financing on a self-sustained basis. The public authorities expect a progressive departure from the internal labour market logic in order to encourage employee mobility and the development of new careers patterns, including the creation of new enterprises. To that effect, changes in the regulatory environment also aim at creating a corporate governance system based on transparent and explicit rules, imposing clear principles of accountability vis-à-vis all stakeholders. The expectation is that it should subsequently be conducive to the kind of flexibility and openness among the different parties in terms of exchange of knowledge and technology that are perceived to be the hallmark of the successful US innovation system. In such an environment, a strategy of risk management could be put in place. This, in turn, would reinforce the innovative drive in large companies and start-ups.

Japanese authorities consider that the economic reforms cannot be successful if they do not proceed alongside the transformation of the tertiary education system. Since the mid-1980s but with sharp acceleration in the last ten years, the shift towards a more result and performance-oriented research environment is undisputed. Competition is strongly emphasized, with the careers of young researchers (and sometimes their professors as well) linked to concrete achievement in their research. Access to funding is becoming more difficult even for institutions such as Tokyo University, where (until recently) the research budget used to be casually rolled over. This creates an environment where commercialization of research and university spin-offs or USOs – defined as newly founded companies that have benefited from management, human resource, capital, knowledge and technology from a university or universities at the time of foundation – emerge as a natural outcome. Thus, the trend is for academic institutions to encourage their academic staff and young researchers to involve themselves in collaborative research projects with private companies in Japan and abroad. The percentage of faculty members and researchers eager to found a company remains very small. However, and considering the high number of university researchers and the fact that more than 50 per cent of basic research is undertaken in universities in Japan (Kondo, 2004), the potential to create USOs remains high. There are nearly 2.5 times as many university-based researchers in Japan as there are in Germany, and more even than in the United States (Ministry of Education, Science & Technology, 2002). The pace of founding spin-offs is accelerating: from 20 to 30 a year in the 1990s to more than 100 a year since the year 2000 (METI, 2006). Many founders are faculty members (professors and senior researchers) – a situation different from Germany, where most are researchers around the age of 30 (Kondo, 2004). This is hardly surprising in a country where the most successful founders of new companies – and especially in manufacturing – are generally experienced managers in their 40s with a long career base in large companies, and often co-owners of patents. These entrepreneurs seek to establish equal footing relationships with their former companies when entering into fields complementing their activities; they also tend to aim towards a fast initial public offering (IPO) of their venture.

The same pattern of creation seems to be observed in the case of USOs in Japan. So far, Japanese university professors have seldom been given the opportunity to acquire the mix of technical and managerial expertise required to create a business. Universities have long conducted research conjointly with companies or have been commissioned by them. However, relations have tended to be of an arm's length nature. Professors were never involved in the commercialization process and were not given the chance to acquire knowledge on business policy, strategy, marketing and human resources management (HRM). Deregulation and the change of legal status of the national universities have changed the rules of the game in this respect, as illustrated by Taplin in this current collection.

The creation of Technology Licensing Organizations (TLOs) inside the universities aims at the adoption of an active intellectual property strategy. 36 TLOs have been registered at the Ministry of Education Science and Technology (Kachi Sogo Kenkyujo, 2006). Professors are now given more opportunities to be involved in actual business, and with the gradual evolution of their mindset (and that of society concerning the proper role of university and their staff in society) a growing number of them are likely to take their chance in the future. Similar to the case of the former managers becoming entrepreneurs it is likely that the status of professor will increase their credibility vis-à-vis business partners and clients. Consequently, one can expect most USOs to be created by senior academic staff.

As for the younger researchers, time does not seem to be ripe yet for significant numbers of those aged between 28 and 35 to launch their venture in manufacturing. Rather, changes will probably come indirectly. Many of these researchers are involved as junior staff in the spin-offs created by their professors and senior researchers. Over time, this involvement should lead to an increase of spin-offs created by younger people. Already, entrepreneurial companies attract an increasing number of young Ph.D. graduates in natural science from large universities. They often prefer to start their career in a new company rather than in a large established one, despite the career risk it may entail and the lower salary. In a second stage of their career a number of them can be expected to create their own company after having acquired enough business experience. As of March 2006, approximately 45 per cent of Japanese university spin-offs were in manufacturing, 37.5 per cent in bioscience and 30 per cent in information technology-related business (Kachi Sogo Kenkyujo, 2006). For the time being their direct and indirect economic impact is small. By March 2006, 1,503 university spin-offs had been founded in Japan, of which 845 resulted from successful research conducted in universities. 658 still have close relationships and 187 have pursued common research with their university of origin. Their direct and indirect economic impact remains limited. By March 2006, they employed 16,383 people directly and 25,858 in total (Kachi Sogo Kenkyujo, 2006). They have on average a 198 million yen sales turnover and 10.9 employees. As of March 2006, the number of IPOs comprised 16 companies, nine of them established in the Kanto (Tokyo and its vicinity) areas: ten in biotechnology, and six in the information technology (IT) business (Kachi Sogo Kenkyujo, 2006). Thus, they are still very small companies. Their creation and development remain plagued with problems of management and corporate governance. A number of regulatory issues have to be tackled thoroughly by public authorities in order to create a sound environment in terms of risk management. Moreover, the debate on the legitimacy, scope and proper objectives of business activities of universities is not yet over. Without clear rules underlying their philosophy in this regard they run the risk of losing their focus, blurring their image and reputation. Conflicts of interest remain a danger, as is the development of a short-term mentality hindering their long-term objectives of research and teaching.

Research Propositions

Against the background described thus far, this study advances the following propositions:

- Although Japanese culture remains overall unsupportive of entrepreneurship, a strong entrepreneurial subculture is slowly emerging in both the business community at large and in universities as a result of education policy changes and exposure to foreign values, leading to changing values in the younger generations.

- Career mobility back and forth from academia to the business world is bound to increase gradually thanks to emerging social acceptance; however, dramatic changes are unlikely to occur.
- Entrepreneurial initiatives will not only emerge from national universities and in the Tokyo area; regional universities can be expected to develop new businesses in collaboration with local public authorities.
- The venture capital industry in Japan will continue to develop, but is unlikely to play a significant role in the foreseeable future.

Context: Emerging Trends in the Japanese Innovation System

Over the past ten years, the rate of mortality of companies in Japan has been higher than that of the creation of new companies (METI, 2005). This trend has to be considered in the context of the restructuring of the Japanese industrial structure. Starting in the 1990s, large companies revamped their supply chain management drastically with severe consequences for the long-term inter-company relationships they had cultivated with small and medium sized companies (SMEs). This together with the subsequent globalization of the economy has altered the economic basis of the companies that had not developed any intrinsic competitive advantage from technology, expertise in production or logistics and marketing. Japan is still a country where the share of SMEs in terms of number and employment is higher than in other developed economies (METI, 2005). But, their economic weight has declined with the increasing number of those unable to compete in the new international division of labour since the beginning of the 1990s.

The Role of SMEs in Japanese Innovation

Despite the massive presence of SMEs in the Japanese economy it always seemed to be a paradox that Japan fared so badly in terms of entrepreneurial drive (Global Entrepreneurship Monitor, 2006). The fact that a majority of new entrepreneurs are people aged over 40, often creating companies out of economic necessity, added to the rather negative image of start-up creation in Japan. However, this does not mean that there are no entrepreneurs in Japan. In fact, many entrepreneurial start-ups have been formed during the last 20 years, despite sluggish economic growth. The most successful cases of the 1980s are companies such as Uniqlo, Autobac 7, Tsutaya, Recruit, Pia or Pasona, while the Internet boom has seen the emergence of Softbank, Rakuten and Manex, among others. However, and with the notable exception of the video game-related software sector where Japan is the world leader (as explained in the study by Storz in this current collection), they have mostly emerged in retail, personnel and other service industries. They have created employment, but most of them of a relatively low level of expertise. Moreover, these companies are mainly active in the Japanese market with little ambition to compete internationally on a large scale.

The emergence of start-ups in the service industries reflects the growing weight of services in the Japanese economy. Moreover, the industrial structure did not encourage the creation of entrepreneurial Schumpeterian types of companies. Now, more career mobility makes acceptance of risk easier in the mind of the younger generation. Previous entrepreneurial booms in history show that there is no cultural atavism making Japanese people less entrepreneurial than the others, if the right conditions exist. Therefore, it can be convincingly argued that the entrepreneurial trend will get stronger in the years to come (Maeda, 2004). Easier access to capital, managerial and technical human resources creates a more auspicious environment for the kind of companies the Japanese government would prefer to see emerging. Whenever manufacturing is located in a country, it seems to be the strong belief of Japanese authorities that innovation always follows. Prowess in the service industries is not enough to maintain the country at the top. Japan has to keep a strong manufacturing basis driven by high technology if, in future, it is to compete with China, India and the United States. Consequently, the emergence of companies in knowledge industries with technological and production expertise is expected. At the same time, these companies should be able to build up what Japanese companies often neglected until recently: a global branding strategy not limited to sales and marketing but implying expertise in management of intangible assets such as legal affairs (intellectual property rights, patents), marketing and finance, as detailed by Taplin in this current collection.

So far, the hopes of the authorities have been largely disappointed. It has remained all together difficult to open up new industries by start-ups. Japan is characterized by a low rate in the creation of Silicon Valley-type fast-growing, high technology entrepreneurial companies in manufacturing. In 2005, only 960 companies were listed on the JASDAQ compared to 3,600 on the NASDAQ. Most of them were concentrated in mature industries such as machinery wholesale and retail (Kachi Sogo Kenkyujo, 2006). A number of well performing high technology firms have been created in the last 15 years, but these are confined to specialized niche markets in fields such as robotics, biotechnology and new materials. Some of them are world leaders in their field, but none of them has yet emerged on a grand scale to succeed Honda, Kyocera, Rohm and Sony, the leading companies of the latest manufacturing start-ups boom just after the Second World War. More than niche market players, Japan is waiting for newcomers that can compete head on in world markets with companies that – in the Japanese psyche – continue to symbolize success in a global economy: Genentech, Cisco Systems, Sun Microsystems, Dell Computers, Hewlett-Packard – all US companies and all former USOs.

The Role of Large Firms in Japanese Innovation

In large Japanese manufacturing companies, innovation and product development capabilities remain high. Japan is at the top for R&D expenses per GDP and number two in terms of patents owned by her companies (*R&D Magazine*, 2006). However, Japanese companies face problems related to product architecture and organizational capability, precisely in industries where Japanese strength is traditionally concentrated: audiovisual, optical and photographic equipment, machine tools, machinery and transportation. Those industries require a high number of single parts whose coordinated development and production is important for the performance of the final product. Japanese industrial and governance structure has been remarkably well adapted to deal with these kinds of products during the last

50 years. Besides the production and logistics expertise they require, they also share common properties as the product development is characterized by incremental type of R&D innovation and the importance of the process innovation (Nobeoka, 2006).

However, the emergence of modular design makes production integration easier for competitors to enter the market (Williamson & Zeng, 2007). Moreover, product commoditization results from function-based customer value as digital technologies easily go beyond basic customer needs. There is no business merit to add more pixels to a camera if average customers do not ask for it and, anyway, would not see the difference because the changes have marginal visible effects (Nobeoka, 2006) In such conditions, competition is increasingly based on cost. Chinese companies and others from emerging economies make tremendous progress in manufacturing modularized products. Moreover, they are heavily investing in R&D in new fields such as nanotechnology and new materials and are bound to compete with Japanese companies in the following years, even in high technology industries (*R&D Magazine*, 2006).

To remain competitive and profitable, large Japanese firms can continue to develop new products with integral design, and concentrate on parts and components rather than final assembly, thus creating values for which customers would be willing to pay a premium and so compensate for the extra costs. That is the strategy followed successfully so far by companies such as Canon, Matsushita, and Sharp, among others. However, both large companies and public authorities in Japan recognize that the necessary breakthrough innovation to create and keep competitive advantages and create new ones in new industries cannot come only from large companies. It also implies a departure from the mainstream way of thinking about innovation. Japanese companies succeeded because of a relentless drive for higher and higher economic efficiency. However, efficiency taken too far can become an impediment to innovation. Breakthrough innovation requires time, money and a speculative mindset.

The Emerging Role of USOs

It is in this connection that the idea has emerged that university spin-offs or USOs could play a decisive role in reforming the Japanese innovation system. Existing companies cannot investigate all the research possibilities. They are under pressure from shareholders in a Japanese corporate governance environment more driven than before by the need for high, short-term profitability. In response, many large corporations downsize their laboratories and shift their resources to more near-future projects. On the other hand, it is the fundamental role of universities to pursue uncertain and speculative types of research. However, because of the constraints built in to the social and economic role of universities, putting the results into practice from inside will always be difficult. To this extent, USOs can fill the gap in pursuing frontier-stretching business projects based on the results of rigorous research into emerging technologies.

Assessment of the Changing Roles of Universities in the Japanese Innovation System

The shift towards an economy driven by knowledge-based companies requires the development of world-class knowledge centres of excellence. Tokyo, Kyoto,

Tsukuba and a few other public and private Japanese universities are high-level centres of education and research. However, Japanese authorities recognize that they are not on a par with the leading US American universities as entrepreneurial engines of development in emerging new industries (Etzkowitz, 2000).

During the 1990s, criticism mounted about the large discrepancy between the resources used by or allocated to Japanese universities and their measurable contribution to practical technology. More than one-third of Japanese researchers were employed in universities and spent about one-fifth of all research funds. However, the entire Japanese university system filled only a very small number of patent applications (Choi, 1999). The drive to promote collaboration between the academic world, business and public administration became very strong after 1995 with examples of collaboration between local public authorities, universities and SMEs in R&D. However, encouragement for the creation of USOs began to be considered as important only after June 2001, when the national government set a target of founding 1,000 spin-offs over the following 3 years (METI, 2002).

The plan succeeded at least in terms of the number of spin-offs created, if not in terms of their performance. In fact, the figure of 1,000 spin-offs is not as important as the commitment to create the appropriate regulatory, business and social environment to allow a widespread revival of the economy. As mentioned previously, this initiative is linked to the revamping of the higher education system in terms of efficiency and effectiveness of resource utilization, and also in terms of its contribution to regional economies. Public authorities make great efforts to encourage university professors and researchers to become more entrepreneurial. Facing tight budgetary constraints, universities need external resources to finance their research. By encouraging activities of technology transfers through TLOs, common research and creation of spin-offs are expected to generate revenue and thus allow the pursuit of basic research independently from public financing.

The problem had been government regulations limiting interactions between professors at national universities and commercial organizations. Traditionally, professors and researchers in national universities have been barred from either taking employment in private companies or from sitting on corporate boards and executive committees. They have not even been permitted to participate in any sort of private venture that might commercialize their research. Such constraints had an impact on SMEs because they have been consistently disadvantaged vis-à-vis larger enterprises in their search for expertise. This has been caused by the lack of a large external labour market for white-collar workers, especially of high technical status. Thus, the restrictions imposed on university personnel further severely limited any contribution universities might have made to the regional economy.

An attempt to allow university staff more freedom was the Law Promoting the Transfer of Research Results from Universities to Private Entities (*Daigaku to Gijutsu Iden Sokushin Ho*), enacted in 1998 with the aim of introducing a more liberal framework. However, it was not immediately clear whether the Ministry of Education (Ministry of Education, Science and Technology since 2001) intended to impose a standard interpretation on all state universities or whether it would allow each institution to interpret the new legislation to suit its own needs. Making national universities independent entities, largely autonomous from the government (since the academic year 2004) indicates a sudden acceleration of the reform process. National universities are now granted control over intellectual property resulting from the work of their research staff, so that knowledge and patented technology can be optimally exploited.

Regionalization Issues

Already 40 years have passed since Jewkes *et al.* (1969) argued for universities to possess public service functions in addition to education and research. Japan has a financial system where money from insurance, savings, and stocks is heavily concentrated in the headquarters of financial institutions in Tokyo. So far, money has been flowing to the regions through public works and state subsidies: that is, as central government expenditure. However, the dire state of Japanese public finance precludes any strong financial back-up from the centre in the future.

From now on, alongside with other types of start-ups, USOs are expected to play an instrumental role in the financial flow of risk capital from Tokyo to the regions. Although IPOs by university spin-offs are an insignificant phenomenon for the time being, once critical mass is reached the financing process is expected to accelerate and diversify. Risk capital funds and business angel initiatives – benefiting from special fiscal treatment and other incentives since the end of the 1990s – are supposed to increasingly originate from the regions themselves. These initiatives should induce a trickle down effect caused by the fact that the majority of USOs tend to establish themselves around the university of origin. In view of their lack of expertise in production processes, marketing and sales, together with shortage of skilled workers, USOs have little choice but to outsource a large part of their activities to small and large local companies.

USOs: Comparing University Performance

In terms of USO rankings, Tokyo University is at the top with 92 spin-offs (as of March 2006) followed by Osaka University with 71 (Kachi Sogo Kenkyujo, 2006). In 2001, Tokyo University established the Advanced Science and Technology Enterprise Corporation to support university venture businesses, dispatching researchers to companies and producing prototypes. Also national, the University of Kyoto (59 spin-offs as of March 2006), launched its International Innovation Center to build corporate relations the same year (*Nikkei Weekly*, 2001). The domination of those three universities – and the strong position of the other national universities as well, with eight out of the top ten in terms of number of spin-offs (as of March 2006) – is not surprising in view of the fact that so far they have received a much larger part of the public research budget than private universities and consequently have had better opportunities to accumulate knowledge.

Regarding venture capital funds, national universities are also the most active. There are venture capital funds in five national universities and in two private ones (Kachi Sogo Kenkyujo, 2006). As a result, national or other public universities have created two-thirds of USOs, while private ones have created one-third. Large universities such as Waseda (75 spin-offs), Keio (50 spin-offs), Ritsumeikan

(30 spin-offs) and Nihon University (29 spin-offs) lead the group among the private universities. The incubator centre created by Keio University in its new campus of Shonan is probably one of the most developed examples of extensive collaboration between universities, local business, administration and networks of connections created by the university's former graduates (Maki, 2006).

Indeed, the dynamism of these large universities reflects the forward-looking strategy of top university administrators, and the large support of the network of alumni in top business positions such in the cases of Keio and Waseda. In some cases, for example those of Doshisha and Ritsumeikan Universities, it is a consequence of the already long-standing links they developed with successful regional clusters in the Kyoto area (Ibata-Arens, 2005). However, the high number of spin-offs from much smaller universities such as Kochi Industrial University in Shikoku (29 spin-offs) or Aizu university (19 spin-offs) in the prefecture of Fukushima – the latter becoming a leader in IT-related spin-offs – demonstrates how both public and private regional universities are becoming increasingly active (Kachi Sogo Kenkyujo, 2006). The Kanto area (Tokyo and its vicinity) took an expected early lead, but now regions such as Chubu, Chugoku, Shikoku and Kyushu are progressing more quickly. Some universities are specializing in one specific field, reflecting their traditional internal strength and/or regional economy comparative advantages. Tokai, Okayama and Hokkaido universities have high-level agronomy and food-related expertise; the Kinki region, including Osaka, Kyoto and Kobe has 34 universities with bio-related departments and around 2,300 bio-related researchers in a region that is the traditional centre of the Japanese pharmaceutical industry (Tamura, 2005).

Trends in USO Ownership and Support

In terms of ownership structure, 68.8 per cent of USOs are joint stock companies (Kachi Sogo Kenkyujo, 2006). Thanks to the changes that were introduced in Japan in 2005 during a large-scale revamp of the country laws governing business organizations, a joint stock company may now be started with a capital as low as one yen (instead of 10 million yen previously), making the total cost of incorporation approximately 200,000 yen in taxes and notarization fees. In the future, it could be expected that university spin-offs will take advantages of the opportunities offered by the Limited Liability Partnership or LLP (*yugen sekinin jigyo kumiai*). LLP may be formed for any purpose (although the purpose must be clearly stated in the partnership agreement and cannot be general), have full limited liability, and be treated as pass-through entities for tax purposes. However, each partner in an LLP must take an active role in the business, so the model is more suitable for joint ventures and small businesses than for companies in which investors plan to take passive roles. Therefore, it may fit well to the needs of USOs where the active and constant involvement of all stakeholders is a key success factor.

According to a survey undertaken in 2004 (*Tsukuba Daigaku*, 2004), almost half (44 per cent) of USOs received a wide range assistance support from their mother university at the time they were founded in matters of mentoring, usage of facilities, contacts with customers and help to secure loans. Afterwards, more than half of them continued to maintain extensive relations with the university, centered on technology-related support. In all, 32 per cent started joint research with their mother universities. The second most common relation is technology guidance from mother universities (20 per cent). While 2 per cent of USOs conduct contract research from their mother universities, 8 per cent of them commission research to their mother universities; however, only 7 per cent provide goods and/or services to them.

Future Perspectives and Implications for Management

Changes in the Japanese innovation system require deeper university-company relationships in order to overcoming the enduring mismatch. For a long time, the weak interlocking between Japanese universities and the business world has been based on informal relationships between professors and a number of companies. These served as conduit to jobs for the best students, a relationship profitable for both parties and one that both universities and companies want to pursue.

Subsequent relationships enlarged to encompass more systematic knowledge and technology transfers, consultancy, common and commissioned research, and patent licensing. In the current period for technological development in Japan, the integration of collaboration with universities and their spin-offs into the long-term strategy of companies could be profitable to both parties. Research shows that corporate and academic alliances and collaboration impacts positively on the technical value of a patent (Sapsalis, 2007). In the future growth and development of USOs there is a need for alliances with large companies in terms of R&D, but also in finance, human capital procurement and market development. Because of the speed of market and technology changes, technological achievements by universities need to be commercialized rapidly. There is also a strong need to develop effective valorization of the patents. This requires an expertise in general management (accounting, HRM, marketing), management of technology and legal affairs (intellectual property rights) that the USOs tend not to possess. There are a growing number of MBA holders in Japan, who graduated from the best American and European business schools. However, they mostly work in large finance institutions and seldom in start-ups. The management of technology that is often taught as a complement to an MBA degree in the United States is just in the first stage of development in Japanese universities.

Conversely, Japanese companies continue to pursue a more open innovation strategy, encompassing both internal and external means. They link their contacts with USOs to their intrapreneurship schemes and to their growing involvement in corporate venture capital (CVC). In a period of increasingly open innovation (that is, the exchange of new ideas with universities, suppliers and outside inventors as increasingly practised by large US companies such as IBM and P&G), Japanese companies have also to find the right mix between a concentration of their R&D resources in target areas and a widening of their range required by emerging technological interfaces and accelerated new technological developments. The opening of R&D activities to the outside also reflects the transformation of the relationships between large companies and their researchers. The evolution of the Japanese labour market towards more mobility and the individualization of career patterns make such relationships based more on short-term 'give-and-take' rather than long-term and as negotiated between loyal and trusting partners. Large companies cannot expect any more that their researchers will stay with them during their whole career and/or that they will not ask for a big share of the profits generated by inventions in the case of commercial success – an emerging expectation illustrated by Taplin in this current collection. Thus, another reason to secure outside knowledge resources is to hedge the risks of losing their internal resources and complement R&D expenses using retained earnings with equity financing in USOs and in other types of start-ups.

On the one hand, universities can afford to consider very risky and uncertain possibilities that companies choose to neglect. On the other hand, once it may be thought that basic university research results could be applicable to commercial use, collaboration with university researchers from the first stage of technology development is required in order for it to be fruitful. This calls for more active human resource interchanges to assess the potential on a continuous basis. Since 2002, professors and researchers have been allowed to serve as executives or advisers for start-ups and other entities that commercialize products or services related to their research, subject to the approval of authorities such as the Ministry of Science and Education through their universities. Among USOs, 7 per cent of their key members are regular faculty staff while another seven per cent work as part-time faculty staff. USOs also accept students from their mother universities as interns or trainees (*Tsukuba Daigaku*, 2004). As of March 2002, 109 faculty members of national universities were involved in outside work as company executives, while 950 were acting as technical advisers (Nikkei Net Interactive, 2002).

As outlined in the introduction to this study, no dramatic increase is expected in the short term and it is likely that mobility will remain limited for a while for most professors and researchers. In fact, the proportion of the Japanese university faculty members who are interested in starting up a company is not higher than 2 to 3 per cent; this compares with 3 to 4 per cent in Germany and 5 to 15 per cent in the United States (Kondo, 2004). However, as the emerging role of USOs becomes recognized as more important than before in the economy, higher social status will probably be given to entrepreneurial activities. Thus, moving back and forth from the university to the business world and/or establishing a company might eventually become a credible and socially acceptable professional model for both experienced researchers and young graduates in addition to managers generally.

By the year 2010, the Japanese government expects 100 university spin-offs to achieve IPO (Maeyama, 2005). However, there is almost no current example of significant success among USOs in Japan. All surveys on USOs highlight financing as the highest entry barrier (METI, 2006). Overall, venture capital markets in Japan are not developed enough: the main venture capital funds are concentrated on established, medium-technology industries and are still very Tokyo-centred. The regional over-the-counter (OTC) markets are still quite small. From 2002 to 2005, 12 university spin-offs listed their stock, ten on the Mothers market in Tokyo, one on the Hercules market in the Osaka Exchange, and one on the Centrex market in Nagoya (Hirao, 2006). Thus, in terms of venture capital there is a need for the general partners to have a high level of knowledge in order to understand precisely the potential of a given technology. In the USA many general partners graduated from a business school. After studying engineering or another natural science they took an MBA, where they studied marketing, strategy and organization theory. In Japan there are extremely few people with such a background. Most general partners come from companies with connections to financial institutions.

As pointed out by Ibata-Arens (2005), a number of successful high technology start-ups created outside of the university were indeed able to overcome the handicap of financing. Some leading companies that created the Kyoto cluster such as Murata, Rohm and Horiba developed through organic growth. They depended on private money to start their business but were able to secure enough cash flow in their early stages of development. However, it would be much more difficult for university spin-offs to achieve such results. In nearly two-thirds of university spin-offs (63 per cent) the number of employees is between one and four at the time of founding. The weighted average is 5.2 employees per company (*Tsukuba Daigaku*, 2004). However, initial capital is relatively large for companies of such a small size. Half of the USOs start with ten to 30 million yen of capital. The weighted average is 77 million yen (Kondo, 2004). These are unlikely to generate cash rapidly in view of the very characteristics of their activities. The sales turnover of USOs is much lower on average than that of SMEs of a similar size. More than 40 per cent of them are losing money, and even among those that reached the stage of IPO, most did not improve their business performance afterwards. After the stock of the 12 companies mentioned previously were listed, it is reported that only three of them had increased their sales and profits: SOIKEN Medical Science Study Laboratory, SOGO Clinical Pharmacology Co., Ltd and Nextech Co., Ltd (Hirao, 2006).

CVCs: A Solution to USO Success?

Thus, another solution is to attract corporate venture capital funds (CVC). In recent years corporate venture capital antennae have been created in leading high technology Japanese companies, and also in the large trading companies (*sogoshosha*). They now send their employees with experience in CVC as general partners in support of venture capital-oriented investments. These people tend to have a scientific background and a good grasp of the technologies utilized in the spin-offs, exactly the expertise required. However, companies often mention that it is difficult to select university partners, make an evaluation of their potential, and manage the relationships beyond personal networks. This is not limited to Japan. Thursby & Thursby (2003) found in the USA that 46 per cent of all academic inventions licensed failed. University technologies are early stage inventions, in many cases no more than a proof of concept at the time of licence. Moreover, no strategic fit is possible with many university ventures because they tend to focus on a single research theme, although the development of almost all new products requires the amalgamation of complex technologies and the securing of several patents. USOs typically develop a very specific technology and few substantial relationships with other laboratories. As the management of the process of putting technologies together is very complex in a university environment, many companies think that the best option is often to limit the relationships to common research and licensing agreements without trying to create a company (Maeyama, 2005).

To facilitate this process, the Japanese government promotes matching funds (*Daigakuhatsu Jigyo Soshutsu Jitsuyoka Kenkyu Kaihatsu Jigyo*) that would pool resources in management and legal affairs. They regroup experts from public research organizations, public and private venture capital, national and local business associations and lawyers associations who are dispatched to support the spin-offs. In industrial clusters such as Tama and Ota, efforts are made to help them to develop a network of relationships locally. Moreover, spin-offs from universities well connected with the business world (such as Waseda and Keio) can utilize their vast network of contacts and access coaching and mentoring services provided by the alumni (Maki, 2006).

Universities all over the world are increasingly eager to protect their inventions and to license their patents. Japanese universities are no exception to this trend. The areas where USOs possess advantages are those where intellectual property rights can be well protected in the form of patent rights possessing high value, or in other forms. Otherwise, existing companies have competitive advantages over USOs because they have marketing, production capabilities, and the R&D and legal expertise to circumvent existing patents. In the case of bioscience it is possible for universities to protect themselves because of the well-established system for managing intellectual property, including patents. The fact that more than 80 per cent of researchers of health science (including medicine and pharmaceutical science) in Japan are university researchers explains why bioscience-related spin-offs account for more than 35 per cent of the total (Kachi Sogo Kenkyujo, 2006).

In principle, access to information is open in universities. Therefore, when a university has started common research with a company, or when a start-up exists, the protection of confidential information is a very important matter. It is disclosed to people and organizations involved during the processes of market potential assessment, business planning and fundraising. However, not all information can be controlled in any university context because excessive confidentiality would run against the very mission of teaching and propagating research results to society. Most Japanese USOs keep close relationships with their original university, mostly of a technological and human resource nature (*Tsukuba Daigaku*, 2004).

Unlike their US counterparts, Japanese universities themselves and their spin-offs currently lack sufficient expertise to manage most aspects of disclosure. The exchange of information in Japanese academic societies is traditionally very informal. Moreover, research in universities is by definition continuous and assumes exchanges with parties outside the universities that are not limited to the joint research process. Consequently, dividing the actions in the universities from those in the spin-off could be difficult to achieve. As a result, and given the relative lack of transparency of corporate governance systems in Japan and the loose control of transactions on the stock exchanges, USOs could be involved in problems stock speculation or insider trading (Hirao, 2006). In this case, the perception of reliability and sense of social responsibility among universities could greatly suffer with detrimental consequences for themselves, their students and for Japanese society generally.

Conclusions: Emerging Types of USOs in Japan

The examples presented in this study indicate that there is, indeed, a growing entrepreneurial subculture in Japan. Moreover, this subculture is concentrated not only in Tokyo; it is emerging as a dynamic force in several regional organizations. Japanese political and business leaders are starting to implement the necessary policy changes and it can be said that, indeed, entrepreneurial spirit in Japanese universities has stronger roots now than before the 1990s.

The 'Silicon Valley Model' may remain alien to the Japanese business and innovation system. Venture capital has not yet reached a stage of maturity. On the one hand, there is very little seed capital around, as the preference among Japanese venture capitalists remains to invest in companies that already have a reliable track record. This eliminates the large majority of USOs from the start. On the other hand, Japanese entrepreneurs active in university spin-offs and in the other start-ups, still appear to have difficulties accepting the interventionism of venture capital funds in the management of their companies.

Meanwhile, the business environment in Japan is witnessing a growing and deeper collaboration between private and public organizations. Despite a small number of widely advertised examples, the idea that 'inventions' are resulting from the efforts and talent of one individual is unlikely to take root: for an emphasis on collective efforts and rewards is bound to remain the dominant norm. Likewise, and although intergenerational power and authority relations are changing, it remains unlikely that a large group of young mavericks will emerge suddenly and challenge the existing order.

In such an environment, one can expect to witness the emergence of different types of university spin-offs or USOs. The fast-growth type that targets world markets is bound to remain a minority. USOs are likely to increasingly develop through external growth (for example, mergers and acquisitions (M&As) and joint ventures sourced from Japan and abroad), accept venture capital funds, and acquire management and legal expertise in order to protect their intellectual property rights (Maeyama, 2005). This current study has highlighted how Japanese USOs aiming at IPO within five years of foundation account for 30 per cent of the national total while those aiming for an IPO within ten years of foundation comprise another 33 per cent. Overall, about two-thirds (63 per cent) of USOs in Japan target an initial public offering within ten years of foundation, a significantly higher percentage than (for example) in Germany (Kondo, 2004). This higher expectation is in line with the higher expertise and status of the spin-offs managers in Japan compared to what is the case in Germany. However, the probability of achieving the government objective of having 100 IPOs by the year 2010 must be assessed cautiously in view of the current mediocre performance of the IPO cases in Japan.

Growth will depend on the development of the risk capital industry (including CVC) and the progress in management practice in the ventures. A larger number of spin-offs are likely to remain small companies targeting niche markets. To assure a steady growth they will (in the main) continue to rely on collecting funds from their founders, their family members and their friends. For the time being, a significant part of the capital comes from the founders, despite the fairly high amount of money required. The USOs, more than three-quarters of whose initial capital is provided by founders, occupy two-thirds (64 per cent) of the total (Kondo, 2004). A third type will develop to respond to the growing willingness of young Japanese people to create non-profit organizations (NPO). With the research back-up of universities, it should be possible to create a larger number of ventures fulfilling social needs without having the ambition to develop products or services which are commercially profitable. This trend may be already present in the recent increase in the number of non-technical university spin-offs, for example in the education field (Kachi Sogo Kenkyujo, 2006).

A fourth type is the research/education promotion type. A USO of this type can be created in order to conduct applied research that is difficult on campus, to supply research materials to campus researchers, to provide concrete research opportunities to students, or to provide both an applied research opportunity and financial assistance to doctoral students. This seems to be the case already in Japan, where currently nearly half (47 per cent) of the staff in spin-offs are made up of students, of which nearly half again are Masters students (*Tsukuba Daigaku*, 2004).

Looking ahead, one problem will be that many spin-offs may not appear to fit into any distinct category, making the task of potential financial backers more difficult. Investors may start to invest in spin-offs that do not ultimately succeed because of an unclear strategy and lack of basic expertise. This may reinforce the caution that financial institutions and companies already appear to express vis-à-vis USOs in Japan. The call is thus for the universities themselves rapidly to develop the necessary expertise to select among the large number of projects underway and target those suitable for establishing a venture company. Furthermore, they need to train their researchers so that they can turn their research ideas into viable business plans more effectively.

References

Choi, J. (1999) Japan educational system heads for reform, *Japan Economic Institute Report*, No. 46 A, 10 December, pp. 9–12.

Etzkowitz, H., Webster, A., Gebhardt, C. & Cantisano Terra, B. R. (2000) The future of university and the university of the future: evolution of ivory tower to entrepreneurial paradigm, *Research Policy*, 29(2), pp. 313–330.

Global Entrepreneurship Monitor (2006) Available at http//:www.gemconsortium.org/download/1154207911515/GEM_2006_Report.pdf (accessed 27 December, 2006).

Hirao, S. (2006) Risk managements for university spin-off ventures, *Report on the International Patent Licensing Seminar 2006*, National Center for Industrial Property Information and Training (INPIT), Tokyo, pp. 130–136.

Kachi Sogo Kenkyujo (Value Management Institute, Inc.) (2006) Daigaku hatsu venture ni kansuru kiso chosa jisshi hokokusho (Basic survey on university spin-offs: report of the empirical results) (Tokyo: Kachi Sogo Kenkyujo) (in Japanese).

Ibata-Arens, K. (2005) *Innovation and Entrepreneurship in Japan* (Cambridge: Cambridge University Press).

Jewkes, J., Sawers, D. & Stillerman, R. (1969) *The Sources of Invention* (London: Macmillan).

Kondo, M. (2004) University spin-offs in Japan, from university–industry collaboration to university–industry crossover, *Asia-Pacific Tech Monitor*, March–April, pp. 24–38.

Maeda, N. (2004) Japanese innovation system restructuring with high-tech start-ups. Paper for SPRIE, a joint research project of A/PARC at the Stanford Japan Center Discussion Center DP 2004-002-E.

Maeyama, A. (2005) Academic start-ups face hurdles, *Nikkei Weekly*, 19 September, p. 13.

Maki, K. (2006) Entrepreneurship in Japan. Keio University/SIVEL, Graduate School of Media and Governance, Taipei, APEC TIC 100 Conference, 7 August.

METI (Ministry of Economy, Trade and Industry) (2002) *Tsusho Hakusho* (White Book on Trade) (Tokyo: METI) (in Japanese).

METI (Ministry of Economy, Trade and Industry) (2005) *Chusho Kigyo Hakusho* (White Book on Small and Medium Sized Enterprises) (Tokyo: METI) (in Japanese).

METI (Ministry of Economy, Trade and Industry (2006) *Tsusho Hakusho* (White Book on Trade) (Tokyo: METI) (in Japanese).

Ministry of Education, Science and Technology (2002) *Kagaku Gijutsu Yoran* (Science and Technology, Digest Book) (Tokyo: Ministry of Education, Science and Technology) (in Japanese).

Nikkei Weekly (2001) Academia builds high technology ties with private sector companies, 22 October, p. 3.

Nishizawa, A. (2006) Risk management for university spin-off ventures. Panel discussion at the International Patent Licensing Seminar, Tokyo, 23–25 January, pp. 130–136.

Noboeka, K. (2006) The influence of product architecture on the competitiveness of the Japanese manufacturers. Paper presented at the 5th Asia Academy of Management Conference, Tokyo, 3 and 4 October.

R&D Magazine (2006) Global R&D Report 2007, September, pp. 7–8.

Sapsalis, E. (2007) The institutional sources of know-how and the value of academic patents, *Economics of Innovation and New Technology*, 16(29), pp. 139–157.

Tamura, M. (2005) Biotech blastoff, *The Japan Journal*, March, pp. 24–25.

Thursby, J. G. & Thursby, M. C. (2003) Industry perspectives on licensing university technologies: sources and problems, *Industry and Higher Education*, 15(4), pp. 289–294.

Tsukamoto, Y. (2005) Present state and issues of the industrial cluster policy of Japan. Available at Research Institute of Economy, Trade and Industry website, http://www.rieti.go.jp (accessed 20 December 2006).

Tsukuba Daigaku (2004) *Daigaku Hatsu Venture no Kadai to Suishin Hosaku ni Kansuru Chosa Kenkyu*, (Research Survey on Issues and Evolution of the University Spin-offs) (Tsukuba: Tsukuba Daigaku Sangyo Liaison Kyodo Kenkyu Center) (in Japanese).

Williamson, P. & Zeng, M. (2007) *Dragons at your Door* (Cambridge, MA: Harvard: Harvard Business School Press).

Emerging Patterns and Enduring Myths of Innovation in Japan: Concluding Thoughts

KEITH JACKSON & PHILIPPE DEBROUX

Introduction

The focus for our discussion in this collection has been on the established and emerging processes and structures of innovation in Japan. The questions addressed by the contributors to this special issue serve (we believe) to illustrate some of the 'emerging patterns' of innovation in Japan. In addition, they motivate us to reconsider some of the 'enduring myths'. In conclusion we can ask:

- What does this varied collection of insights tell us about what is enduring and what is changing in the context of innovation in Japan?
- What will – or should – Japanese-style innovation look like in the future?

Enduring Myths

To invoke a well-worn and still worthy phrase: in order to understand the present, we need to understand the past. We contend that a number of past perspectives on innovation in Japan have generated a series of enduring myths. One of these relates to its 'uniqueness'. To what extent can the problems and solutions of innovation in Japan be truly or even usefully described as being 'unique'? In the introductory chapter to this collection we highlighted some of the achievements of Japanese-style innovation: for example, as demonstrated by elements of the Toyota Production System (TPS) and the 'home-grown' management processes, structures and cultures that can serve to make Japanese organizations 'unique' by global comparison. This approach has an impressive track record (cf. Ouchi, 1981; Takeuchi & Nonaka, 2004). The myth – or insight – of 'uniqueness' in respect of Japanese-style innovation will endure for as long as there is an audience willing to sustain it.

However, making claims to 'uniqueness' has a tendency to provoke envy or scepticism and cause some business commentators to trade in generalizations. For example, in the mid-1990s commentators adopting a North American perspective could claim that 'innovation in Japan' (that is, perceived as being subsidized and protected by statist intervention and other socio-political structures) was all about 'going for singles' (Herbig, 1995). In contrast, the United States – being more open to markets and to the free transfer of people and of creative ideas – demonstrated a propensity in innovation management to 'go for home runs'. We should assume that Herbig was careful to choose metaphors from a sport that arouses such passion on both sides of the Pacific. Claims that seek to emphasize collective uniqueness – and, by implied attribution, difference in relation to other collectives – serve to reaffirm the generalizations expressed by comparisons of 'national cultures'. Such claims invoke neat images drawn of 'the Japanese' as being generally more collectivistic and risk averse in their business thinking than their counterparts in the United States of America (cf. Hofstede, 1993, 2001). We should remember that by applying such generalizations we are talking about a combined population of around 370 million people – or over a quarter of the current population of China. How many people does it take to sustain a national myth, and for how long?

A commitment to such generalizations feeds naturally into the imperfect methodologies of many surveys designed to produce comparative (and perhaps competitive) rankings. Rankings continue to claim excited reference in Japanese society: Which is the 'highest ranking' university to graduate from? Which is the 'most famous' company to work for? The answers when given may serve more to reinforce a myth rather than prompt people critically to reassess the future value of their current situation. As an illustration, a recent survey for the IMD World Competitiveness Yearbook (Garelli, 2005) ranked Japan as 59th out of 60 in the category 'entrepreneurship of managers'. In terms of language skills (that is, skills relevant to meeting the needs of international enterprise), Japanese managers are rock bottom at 60.Their US counterparts come in at 51 in this category; however, and given the persistence of English as the global business language of choice, this offers scant consolation from a Japanese perspective.

Against the background of our current discussion, perhaps one of the most telling comparisons in the IMD (Garelli, 2005) survey is that comparing the relative 'openness' of 'national cultures' to 'foreign ideas'. In this category, Hong Kong comes top, regional competitors Taiwan and South Korea are 4th and

7th respectively; Japan lies 14th, while the USA come in at 25 of the 60 economies surveyed. What are such rankings to tell us about the enduring (self-) images of the insular Japanese manager? How might they be used to inform the nature of recruitment and training policies in Japanese and, indeed, international organizations? What do they say about the role of the Japanese government and the state of the national education system that supports these outcomes?

Adopting a more individualized perspective, what are new generations of Japanese managers to make of such comparisons: that is, the type of manager, researcher, entrepreneur and inventor who has figured so prominently in the contributions to this current collection? For a start they could question the assumptions upon which such 'global surveys' are based, together with the bias inherent in the methodologies they use. They might shrug and conclude that Japanese-style inventiveness and flexibility in business have been underestimated by 'outsiders' (*gaijin*) so routinely in the past; and that Japanese industry and society still generate the second largest economy in the world while sustaining one of the most literate, numerate and equitably developed societies ever to exist in the history of collective human endeavour (cf. World Bank, 2006).

Emerging Patterns of Innovation: The Human Dimension

Overall, the studies presented in this collection serve to connect with and build on a growing corpus of Asia-Pacific innovation research. More specifically they combine to help us trace some of the more significant and emerging changes in the patterns of innovation in Japan, illustrating this design with examples from diverse industries and business sectors, and developing arguments from a number of internal and external perspectives.

For example, tuning in to the emerging patterns of interaction illustrated in this current collection provides further evidence of the potential that exists already in Japan to develop what scholars in another national context have termed 'triple helix cooperation' between state, academic and industrial institutions (cf. Nowak & Arogyaswamy, 2007). Correspondingly, local government and business leaders in Japan are becoming increasingly and actively aware of the opportunity they have to look beyond central government and learn from other countries and regions (for example, the Silicon Valley model in California) and work out systematically how such policies might be adapted to future innovation-oriented social and business environments in Japan (cf. Nishizawa, 2007). These and other actors reconsider what lessons can be learnt from the 'enduring myths' of innovation in Japan, and the extent to which reference to these myths might serve to help or hinder the emergence of more sustainable patterns of nationally-based innovation in an increasingly globalized business environment.

Talent

Ultimately, it comes down to the people involved. For, implicit in each of these conversations is the primacy of the human element: or, in emerging human resources management (HRM) terms, the 'talent' or employees who express, among other qualities, a critical and strategic mind and entrepreneurial instincts (cf. Michaels *et al.*, 2001; Ohmae, 2001). The focus on 'talent' is one form of HRM response in Japan – as elsewhere – to the turbulence generated by increasing change and uncertainty in the global business environment and, by extension, in global markets for labour (cf. Kono & Clegg, 2001; Debroux, 2003a). The shifts in domestic Japanese markets for labour and career development provide one example of how established or standard patterns of HRM architecture, policy and practice in large Japanese organizations are converging with those of their non-Japanese competitors: for example, in terms of developing new HRM systems that might accommodate more effectively the emergent significance of individual talent (cf. Benson & Debroux, 2004; Jackson, 2005).

In respect of striving towards more effective management of innovation, this shift in HRM focus is as true for Japan as it is for any other ambitious nation, as emphasized by Simon (2007) in his concluding comments to an *Asia Pacific Business Review* special issue titled *Global R&D in China*. In this context, Simon and his colleagues recognize the 'strategic role of talent'. We agree. For, developing a 'global' HRM perspective on innovation prompts us to tune in to the scholarly and insightful conversations converging in this collection and infer from them the voice of an emerging generation of Japanese managers, academics, entrepreneurs and researchers: in short, of innovators. These innovators are developing a routine vision of innovation that entails modelling and re-modelling career decisions, each expressing a willingness to take the type of business and career risks required to make innovation in Japan real and sustainable (cf. Jackson & Tomioka, 2004).

Innovation in Japan: Reassessing Past Myths

Myths are sustained through the combined efforts, interests and (under-) achievements of people in a particular place and time. In confronting such issues, management scholars and practitioners can seek out insights developed by historians and social anthropologists. As an illustration, Kurita explains how 'the *essence* of Japanese identity is most plainly manifested in an attitude toward nature and the four seasons. It is an attitude that can be summed up by the phrase *setsu-getsu-ka*, snow, moon and flowers' (Kurita, 1987: 13; emphasis in original). By this token, we might interpret that being Japanese is akin to recognizing that nature can be everywhere improved upon but never surpassed or ultimately brought under control: paraphrasing from Dunn (2005), *being* Japanese implies adopting a distinctively 'patterned' identity. From an HRM perspective and in the context of organizations, this shared identity might be interpreted (fatalistically, perhaps) as *unmei kyodotai* or 'a community of fate' (Debroux, 2003b; Koike, 1995).

The potential for combining reference to enduring national myths with an irrepressible desire to create new patterns of national and organizational identity and innovation is expressed in the *ukiyo-ē* ('floating world') prints of Hokusai, an artist of pre-Meiji Restoration (that is, pre-industrial) Japan. Even today, Hokusai's prints evoke many of the traditional and mythical patterns of what it means to be Japanese. To cite from a modern source, this master of craft and invention continues (after death) to draw on the nationally defined 'DNA' of Japanese imagination and design (cf. JETRO, 2006).

As a Japanese artist and artisan, Hokusai became legendary for combining eastern and western styles and techniques, balancing his use of these to serve both domestic demand and to develop European/western markets (cf. Forrer, 2004). In reflecting on Hokusai's work, it is possible to perceive contemporary echoes of Toyota cars: that is, as the products of a master manufacturer, designing to develop and satisfy domestic demand while simultaneously achieving spectacular success in global markets; surpassing all rivals in terms of sales volume and, it might be argued (by Toyota managers, probably) of innovation.

The Future of Innovation in Japan

Drawing on the insights presented in this current collection, it is possible to stay reasonably optimistic about the future of Japan as an innovative country. The third basic plan in science and technology launched in March 2006 focuses on the commercialization of technologies, and on public education in the scientific and social potential effect of the discoveries of the last decade. Its key objectives seem to be realistic in order to achieve significant results in science, technology and sustainable development. Focus is put on patents and their management, and on the financing of research through competition-driven subsidies, while keeping a national evaluation system. It is a holistic approach also putting emphasis on promotion of a culture of research among young Japanese people, female scientists and on development of contacts with foreign researchers. It encourages experienced researchers to reinforce the collaboration with government, universities and industry in specific areas such as anti-cancer measures, nanotechnology and related materials, and industrial robots.

For now, the Silicon Valley cluster model for innovation may remain largely alien to the Japanese scientific and business environment. It is unlikely that the same entrepreneurial drive based on innovative start-ups will suddenly emerge in Japan. However, at least successive Japanese governments and private organizations appear now to share a commitment towards keeping Japan a competitive country in the world. Issues of political loyalty and divisive interest exist in Japan, as elsewhere. However, remembering the lesson of the semi-conductor industry failure – as detailed by Okada in this current collection – efforts are being made to encourage collaboration, association and collective innovation in involving a wider array of players. By doing so, the national innovative system becomes more flexible and responsive. Consequently, in their own way and without changing fundamentally their traditional paradigm, what Japanese organizations are trying to achieve in terms of innovative activity may prove to be successful in the long term.

Questions for Future Research

Some emerging patterns of innovation in Japan are now established and they suggest ambitions far beyond hitting only 'singles', as diagnosed provocatively by Herbig in the mid-1990s. Japanese 'home-run' experts in fields as diverse as baseball (Hideki Matsui) and (figuratively speaking) science and technology are now being 'exported' to the USA and to other countries. The key questions that researchers – together with politicians, managers, and other key actors in Japanese innovation – need to address from now on include:

- To what extent do the myths of the past in Japan act as a drag on the potential for innovation now and in the future? To what extent might they act as a spur: as, for example, our invocation of Hokusai's innovative craft is designed to suggest? To what extent are/were these myths accurate anyway?
- Why are some industries and business sectors in Japan moving faster than others in terms of expressing their innovation capability? Why are some policy decisions by central and other tiers of government working more effectively than others? How do the emerging patterns of managed innovation in Japan (for example, in response to government policy initiatives, or policy misadventures) compare to similar initiatives in other national and regional contexts: for example, the Special Economic Zones (SEZs) and Special Administrative Regions (SARs) of China (cf. Sun *et al.*, 2007)?
- How and when are Japanese universities and other non-*kaisha*-based research institutions going to realize more of their full potential in terms of innovation and the commercialization of innovative activities? When is the long-heralded 'internationalization' of education in Japan going to be realized in practice?
- How might Japanese people generally be motivated and empowered to promote entrepreneurship and innovation? To what extent are Japanese people willing and able to share the development of their potential for empowerment and innovation: with each other and with other people: for example, non-Japanese people in Japan and elsewhere?

The history of innovation in Japan – the foundation for the 'national innovation system' of Japan – is one illuminated by a glittering series of success stories. Many of these successes have been planned; many have been improbable, but true – one thinks of Sony, a giant of innovative manufacturing and marketing, created from among post-Second World War ashes by two (now) legendary entrepreneur-inventors, Ibuka and Morita. This series of success stories will continue into the future. The significant strategic questions for all stakeholders involved relate to how the patterns of these future successes will emerge and – beginning now, and even as general business confidence (at the time of writing this in March 2008) appears to wane – to how they will be managed in order to achieve their full potential in terms of both business and human development.

References

Benson, J. & Debroux, P. (2004) Flexible labour markets and individualized employment: the beginnings of a new Japanese HRM system? in: C. Rowley & J. Benson (Eds) *The Management of Human Resources in the Asia Pacific Region: Convergence Reconsidered*, pp. 55–75 (London: Frank Cass).

Debroux, P. (2003a) *Human Resource Management in Japan: Changes and Uncertainties* (Aldershot: Ashgate).

Debroux, P. (2003b) Culture and management in Japan, in: M. Warner (Ed.) *Culture and Management in Asia*, pp. 99–114 (London: RoutledgeCurzon).

Dunn, M. (2005) *Inspired Design: Japan's Traditional Arts* (Easthampton, MA: Five Continents).

Forrer, M. (2004) *Hokusai: Mountains and Water, Flowers and Birds* (London: Prestel).

Garelli, S. (compiled) (2005) *IMD World Competitiveness Yearbook 2005* (Lausanne: IMD).

Herbig, P. A. (1995) *Innovation Japanese Style* (Westport, CN: Quorum).

Hofstede, G. (1993) *Cultures and Organisations: Software of the Mind* (London: Harper Collins).

Hofstede, G. (2001) *Culture's Consequences* (London: Sage).

Jackson, K. (2005) Divergence in global HRM: managing trust and talent in Japan, Plenary paper for the online conference, The Relationship between cultures and International HRM. Available at http://www.dialogin.com (accessed 22 May 2008).

Jackson, K. & Tomioka, M. (2004) *The Changing Face of Japanese Management* (London: Routledge).

JETRO (Japan External Trade Organization) (2006) *DNA of Japanese Design* (Tokyo: JETRO).

Koike, K. (1995) *Nihon no Koyo System* (The Japanese Employment System) (Tokyo: Toyo Keizai Shimposha) (in Japanese).

Kono, T. & Clegg, S. (2001) *Trends in Japanese Management: Continuing Strengths, Current Problems, and Changing Priorities* (New York: St. Martins Press).

Kurita, I. (1987) *Japanese Identity* (Tokyo: Fujitsu Institute of Management).

Michaels, E., Handfield-Jones, H. & Axelrod, B. (2001) *The War for Talent* (Cambridge, MA: Harvard Business School Press).

Nishizawa, A. (2007) University start-up ventures and clustering strategy in Japan, in: R. Taplin (Ed.) *Innovation and Business Partnerships in Japan, Europe and the United States*, pp. 102–131 (Abingdon: Routledge).

Nowak, A. & Arogyaswamy, B. (2007) Innovative practices in Poland: an organising framework and action plans, in: R. Taplin (Ed.) *Innovation and Business Partnerships in Japan, Europe and the United States*, pp. 54–69 (Abingdon: Routledge).

Ohmae, K. (2001) *The Invisible Continent: Four Strategic Imperatives for the New Economy* (London: Nicholas Brearly).

Ouchi, W. G. (1981) *Theory Z: How American Business Can Meet the Japanese Challenge* (Reading, MA: Addison-Wesley).

Simon, D. F. (2007) Whither foreign R&D in China: some concluding thoughts on Chinese Innovation, in: Y. Sun, M. von Zedtwitz & D. F. Simon (Eds) *Global R&D in China*, pp. 471–480 (Abingdon: Routledge) (special issue of *Asia Pacific Business Review*, 13(3)).

Sun, Y., von Zedtwitz, M. & Simon, D. F. (Eds) (2007) *Global R&D in China* (Abingdon: Routledge) (special issue of *Asia Pacific Business Review*, 13(3)).

Takeuchi, H. & Nonaka, I. (2004) *Hitotsubashi on Knowledge Management* (Chichester: Wiley).

World Bank (2006) *World Development Indicators* (Washington, DC: The World Bank).

INDEX

Page numbers in *Italics* represent Tables and page numbers in **Bold** represent Figures

academic societies 171
Advanced Science and Technology Enterprise
Corporation 166
Advanced Semiconductor Research Center
(ASRC) 110
Advanced System-on-a-Chip 109
Advanced System-on-a-Chip Platform
Corporation (ASPLA) 110
aerospace industries 12, 13
Aizu University (Fukushima)167
ambiguity 125
American model of innovation 116
American software industry 37
Ando Electric 103
angel financing 126
Angel tax 144
Angels vs. "Classic" Venture Capitalist 41, *42*
Asia Pacific Business Review 4, 178
Asia-Pacific economies 3
Asia-Pacific innovation research 177
Asian financial crisis 19, 23
Association of Super-Advanced Electronics
Technologies (ASET) 110
Association of University Technology
Managers (AUTM) 79
Atlus 121
Autobac 162
automobile industry 14, 26, 125

Baba, Y. 122
BandaiNamco 121
Banpresto 121
basic science 64
Bayh-Dole 43
Bayh-Dole Act in the United States of America
(1980) 37
Behaviour, strategic 9, 13, 14, 15, 25, 26, 27
benrishi 82, 83, 88
Berkeley University 32

"best-practice" model 133
biopharmaceuticals 32
bioscience 161, 171
biotechnology 31, 37, 115, 117, 129, 159;
industry 46
"Bit Valley" 121
Bit Valley Association (BVA) 128
Bit Valley Party 121
Bush administration 48
business angels 2, 6, 139–56; characteristics
and activities 148; defined 141; model 144;
role 140
business decisions 153
business software 115, 121, 129
business strategy 5

California 48
California Tech 32
Canadian business angels 149
Canon 164
Capcom 121
capital market 119, 126
capitalists 2
Center for Venture Research at the University
of New Hampshire 144
Central Government role 2
central processing unit (CPU) 105
Centrex market in Nagoya 169
chemicals 10
Chief Executive Officers (CEOs) 149, 151
Chief Technology Officer (CTO) 151
China 158, 163
Chinese companies 164
Chinese intellectual property rights officials
87
Chugoku 143, 167
chûken kigyô 121
Cisco Systems 163
Civil Procedure law 90

classic Venture Capitalists (VCs) 40, 41, *42*, 144, 145
Clinical developments in China 48
Code Division Multiple Access (CDMA) cellular phone systems 14
Collaborative Achievement (ASUKA) Project 109
collective invention 131
collectivism 20
collectivistic societies 20
Commerce Law revisions 1–4
Commercialization 34–9, 60
commercialization of innovative activities 180
commercialization of technology 44, 179
commercialization in US universities *36*
community of fate *unmei kyodotai* 178
company alliances 108
company internal coherence 20
Comparative Management Research 25–7
competitiveness 121–3
Confucianism 20
constraint on management resources 66
consumer electronics 117
Contingencies; Strategies and Changes 98
continuous improvement (*kaizen*) 1
conversion 124
"Cool Japan" 115–38
"Cool" Japanese products 121
cooperative learning 9, 95
Core Research Concept Definition 116
"core technology" 64
corporate cultures 22
corporate governance 22; systems 171
corporate innovation management 9
corporate response 3
corporate venture capital (CVC) 168, 170–1, 172
counterfeit goods 87
creative improvisation 131
Critical Path Initiative 45
cross-functional integration 24
cumulative venture capital investment *42*

Day of Invention 82
Debroux, P. 1–7, 148, 157–74, 175–81
Dell Computers 163
design and fashion 23, 25
"design-first" approach 18
Deutsche Bank Research 118
development of semiconductor industry 102
development of technologies 14
digital entertainment 134
Digital Games Research Association 132
Digital Hollywood University 128
digital technologies 164
Disintegration of the Old Structure 102

displacement 127
District Courts; Tokyo or Osaka 89
diversification degree 14
diversification strategies 14
domestic market 2, 22
Doshisha 167
Drucker, P. 154
dynamic cooperation and partnership 4
Dynamic Innovation Patterns 111
dynamic random access memory (DRAM) 93, 94, 100, 101, 102, 103, 104, 105, 108, 111

economic reforms 159
economic stagnation 78
Edo Shogunate 77
Electro-Technical Laboratory 102
electronics 13, 26
Elpida Memory 105, 108
embryonic stem cells 34, 48
emerging patterns 175–81
Employees' Rights to Compensation 5, 79–82
employment growth *38*
enduring myths 175–81
enterprises; small and medium-sized (SMEs) 5, 6, 46, 56, 58, 60, 63, 65, 66, 69–73, 74, 78, 88, 118, 122, 158, 162, 165, 170
entrepreneurial management 154
entrepreneurial start-ups 158
Entrepreneurs 115–38
exchange of information 171
export performance *13*
external knowledge 131; reliance 23
external suppliers integration 25
external technology 24

firearms 77
foreign ideas 176
Fujitsu 105, 108
Fujitsu v. Texas Instruments 89, 90
Fujitsu-Hitachi Plasma Display 108
Fukuoka 143
future innovation 179
future perspectives 168–71
future research 179

game; internal coherence 125
game designers 126
Game Developers Conference in US 131
game software developers 122
game software industry 116, 121–3, 129
game software publisher 121
game software sector 128
game software shops; second-hand 122
Genentech 40, 163
general machinery 10
genesis of game software sector 130

German Bank 117
Germany 118, 160, 169, 172
global change response 2
Global Entrepreneurship Monitor (GEM) 41,
   116, 118, 162
global leadership 25
Global Life Science Industry 31, 32
Global R&D in China 178
globalization 55
government grants 147
government subsidies 112
Grand Panel system 84
granting of rights 86

Hague Agreement Geneva Act 87
hardware makers 128
Harrison, R. 141, 145
Harvard University 6, 32, 48
health-care products 32
health science 171
Hemmert, M. 3, 4, 8–29
Hercules Market in Osaka Exchange 169
heterogeneity 125
Hewlett-Packard 163
Highly Agile Line Concept Advancement
   (HALCA) 110
Hiranuma Plan 143
Hitachi 81, 103, 104, 105, 108
Hitachi-Fujitsu-Sanyo 108
Hitachi-Mitsubishi-Matsushita 108
Hokkaido University 167
Hokusai, K. 178, 179
Honda 163
Honda Motor Company 87
Hong Kong 26, 176
Horiba 170
human dimension 177
human resources: issues 151; management 4,
   18, 24, 177; and venture capitalist funding
   152
Hungary 41
Hyundai Motor 14

i-mode 129
Ibata-Arens, K. 4, 30–53, 46, 144, 159, 170
IBM 168
Illinois governor 48
IMD World Competitiveness Yearbook 176
incubation facilities 150
incubator centre (Keio University) 167
independent core suppliers 18
India 158, 163
individualistic societies 20
Industrial Cluster Pan 143
industrial clusters 171
industrial and governance structure 163

industrial organization 120, 127
Industrial Property Administration 85–8
Industrial Revitalization Law 37
industrial robotics 1
industrial structure 130
industries electronic and microelectronic 10
industries high technology 12
Informal Venture Capital in US 144
information technology-related business 161
information-oriented society 86
infringement 89
initial public offering (IPO) 160
innovation 115–38
innovation capability 180
Innovation Cluster Initiative 46
innovation competition 55
"Innovation and Entrepreneurship"
   (Ibata-Arens) 46
Innovation Management 8–29, *21*, 24
innovation patterns 3, 6
innovation processes 69
innovation system trends 162
innovation-related activities 10, 11
innovative capabilities of companies 96
institution conversion 124
institutional arrangements perspective 96
institutional change modes 124
institutional complementarities 115
institutional innovation in game sector 116
Institutional and Non-institutional
   Contingencies 97
institutions 115–38
intangible assets 5
Intellectual Property (IP) 5, 13, 77–92, 79;
   Basic Law 143; Dispute Processing 86;
   Divisions 78; High Court 83; overseas
   protection 87; production 11; related
   problems 61; rights 172; Strategic Program
   84; Strategy Committee 84; valuing 82–5
Inter-Firm Cooperation 93–114
interfaces between institutions 120
internal technology sourcing 23
international accord 87
International Business Machines (IBM) 105
international competitiveness 18
International Innovation Center 166
International Monetary Fund (IMF) 4
international power 172
international R&D 17
international trade 117
internationalization of education 180
internet boom 162
Intra- and Inter-Firm Interaction 96
intra- and inter-firm cooperation 5
IPS Alpha Technology 108
Ishimaru, K. Dr. 80

J-model of management 125
Jackson, K. 1–7, 175–81
Japan Association for New Business Incubation
    Organizations (JANBO) 39
Japan corporate R&D 10
Japan External Trade Organization (JETRO)
    3, 7
Japan Finance Corporation of Small and
    Medium Size Enterprises (JASME) 56
Japan Patent Office (JPO) 83, 85–8
Japan patents 31
Japanese economy stagnation 8
Japanese exports 3
Japanese government 4
Japanese Innovation System implications 73
Japanese national innovation system 49
Japanese NIS reform 6
Japanese NIS stature 5
Japanese Patent Office 57
Japanese Statistical Bureau 54
Japanese-style economic system 55
Japanese-style future innovation 175
Japanese style of management 124
Japanese-style venture businesses 122
Japanese universities 148
Japan's Competitiveness Nature 117
Japan's exports 117
Japan's Patent Law 81
JASDAC 163
job rotation 126
Johns Hopkins University 32
Joint Research Center for Atom Technology
    (JRCAT) 101
jump factors 31
Just In Time (JIT) inventory management 1

Kagono, T. 129
Kanagawa 143
Kanagawa Prefecture 149
Kanto areas 161, 167
Kanto region 153
Keio University 32, 166, 171
Key Policies 43
Kinki and Hokkaido regions 46
Kinki region 167
Kirihata, T. 146
knowing-doing gap 120, 130
knowledge-based companies 164
knowledge centres of excellence 164
knowledge creating company 131
knowledge exchange 129
knowledge industries 163
knowledge-intensive service sectors 121
Knowledge supporting university spin-offs 147
knowledge and technology transfers 168
Kochi Industrial University in Shikoku 167

Kohashi, R. 129
Koizumi; Prime Minister 82, 84
Koizumi Government 78
Konami 121
Konosuke Masushita 78
Korea corporate R&D 10
Korean economy 22
Korean firms 8–29
Korean management style 22
Korean Ministry of Labor 19
Korean semiconductor companies 103
Korean War 22
Kurita, I. 178
Kyocera 163
Kyoto 143, 164
Kyoto cluster 170
Kyoto University 32, 37, 166
Kyushu 143

labour market 119, 168
labour markets 127
Large Firms role 163
Large Scale integration (LSI) 101
large scale integration processing equipment 110
large trading companies (*sogoshosha*) 170
large-scale integrated circuit (VLSI) 94
Law Promoting Technology Transfer (1998) 36
Law Promoting the Transfer of Research
    Results from Universities to Private Entities
    165
Law on Special Measures for Industrial
    Revitalization 37
Law to Strengthen Industrial Technological
    Capabilities 55
lawyer (*Bengoshi*) 88
leading game software publishers *123*, 133
learning by experimentation 131
learning by searching 131
Life science National Innovation Systems *49*
life science research 32
life sciences 40
lifetime employment 1–4
light emitting diode (LED) semiconductor 81
Limitations of Vertical Cooperation 101
Limited Liability Partnership (LLP) 167
liquid crystal display 14, 80, 105
lithography developments 110
litigation attitudes 5
long-term employees (*shains*) 125
long-term employment 124
"long term orientation" 20
Lundvall, B-A. 95

macro-level strategies 99, 108
Madrid Agreement Concerning the
    International Registration of Marks 88

magnetic random access memory (MRAM) 108
Management of Firm Boundaries 64
management processes 176
management of technology 168
managerial evaluation 23
Managerial strategies 23
Manex 162
manga industry 134
manga tradition 130
manufacturing 161
manufacturing and marketing 74
manufacturing technologies 65
market-oriented mechanisms 111
Mason, C. 141
Massachusetts Institute of Technology (MIT) *34, 36,* 37
matching funds 171
Matsushita 104, 164
Matsushita group 78
mechanical engineering 117
medical devices 32
medium-sized firms 121
MegaChips 101
memory chips 14
mergers and acquisitions (M&As) 104, 172
Michigan University 32
micro-level strategies 99
microelectronics 26
Micron Technologies in US 100
Microsoft Corporation research report 131
middle-up-down management 1
millennial analysis (Porter) 4
Millennium Research for Advanced Information Technology (MIRAI) 110
Ministry of Economy; Trade and Industry (METI) 56, 139, 141, 143, 146, 148
Ministry of Education: Culture; Sports; Science and Technology (MEXT) 141; Science and Technology 165
Missouri 48
Mitsubishi 105
Mitsubishi Research Institute 56
Mitsubishi/Sharp 108
Mitsui Zaibatsu 77
mobile telephone market 122
modular design 164
modularization 111
Modularized cooperation and competition 111
modularized product areas 100
Monsanto 32
Mothers market in Tokyo 169
Motohashi, K. 5, 54–76
Munekuni, Y. 87
Murata 170
myth of "uniqueness" 176

Nagoya 158
Nakamura, S. vs. Nichia Corporation 80
NamcoBandai; publisher 126
nanotechnology 12, 164
NASDAC 163
National Business Incubator Association (NBIA) 39
national cultures 176
National Innovation Systems (NIS) 4, 30–53, 54–76, 115, 180; concept 119
National Institute of Advanced Industrial Science and Technology (AIST) 80, 110
National Institute of Health (NIH) 33
National Policy 42–5
National Policy Developments 45
national system of innovation (NIS) 95
NEC/Matsushita 108
New Energy and Industrial Technology Development Organization (NEDO) 57
new firm start-ups 30
New inter-firm alliances *106*
new materials 164
new software 117
New Structures with Horizontal Cooperation 103
new technology-based firms (NTBF) advance 74
New Technology development 30–53
new technology-based firms (NTBFs) 5, 55
newly established and liquidated firms **118**
Nextech Co. Ltd. 170
Nihaon University 167
Nikkei Newspaper Survey 54
Nintendo 117, 121, 122, 128
Nippon Electric Corporation (NEC) 103, 104, 105
Nishizawa, A., Professor 79
Nissan 129
Non-governmental Agencies 2
non-institutional factors 97
non-profit organizations (NPO) 173
non-technical university spin-offs 173
North American perspective 176
North, D. 95
"not-invented-here" 33
"not-invented-here" syndrome 55
nuclear fuel business 108

Oberländer, C. 3, 4
Office for Harmonization in the Internal Market (OHIM) 88
office machinery/computer industries 13
Office for Promotion of Justice Systems Reform 83
Okada, Y. 5, 93–114, 159, 179
Okayama University 167

Oki Electric 105
Open Source Software 131
optical disc technology 81
Organization for Economic Cooperation and
    Development (OECD) 2, 17, 41, 115, 116,
    117, 118, 119
Organization for Small and Medium Enterprises
    and Regional Innovation in Japan (SMRJ)
    survey 141, 149, 150
organizational form of national universities 55
Orphan Drug Act 44
Osaka 143, 158
Osaka District Court 90
Osaka Prefecture 149
Osaka University 32, 37, 143, 166
Ota 171
outsiders *gaijin* 177
outsourced research and development **58**
Over-The-Counter (OTC) 159
over-the-counter (OTC) markets 169

Paris Convention 78
Pasona 162
patent applications 165
Patent Attorneys (*Benrishi*) 88
Patent Attorneys Law 83
Patent Cooperation Treaty (PCT) 87
Patent Court System 88–91
Patent Courts 82
Patent examination 85
Patent generating universities 32
Patent Law 79–82; reforms 83
Patents 163; management 179; payback 157;
    policy 43; systems 46; validity 89
path dependencies 115
path dependency 116
performance based approaches 24
performance-based compensation 19
performance-oriented research 160
peripheral institutions 133
pharmaceutical industry 14, 167
pharmaceuticals 10, 13
Ph.D. graduates 161
Pia 162
plasticid 117, 132
plasticity 116, 130; Institutional Foundations
    124
"plasticity" concept 6, 124
plasticity displacement 134
plasticity of Innovation Systems 124–9
playstation 122
PlayStation II 105
Playstation3 of Sony 122
Policy developments of USOs 141
policy for science-based new business creation
    47

Porter, M. 3, 4
power semiconductors 108
pre-Meiji Restoration 178
private corporations 147
private education institutes 128
"private equity" investment 39
privatization of Universities 78
"pro-entrepreneurial culture" 5
process innovations 65
process management practices 14
Proctor and Gamble (P&G) 168
product development time 18
professional infrastructure 152
profile of Japanese Venture Capitalists (VCs)
    146
project management 24
pull; push and jump factors 31
push strategies 31

Rakuten 162
rare diseases 44
Recruit 162
regional variations 158
regionalization issues 166
regulatory relaxation 150
Renesas Technology 108
research cooperation system 112
research cooperatives 101
Research and Development (R&D) 10;
    collaboration 24, 54–76, 61–4, **64**, **71**;
    collaboration and environment **63**;
    collaboration of SMEs 56, 73; Collaboration
    Survey 56; content in-house 68–9; corporate
    efforts 13; effect of collaborations 58;
    environmental change 61–4; expenditure 2,
    10, **11**, **12**, 17; expenses 163; external
    collaborations 57; funding 147; in-house 64,
    64, 69; intensities 13, 14; investments 20,
    23; managing 4; personnel 151; private
    incentives 2; productivity and in-house *70*;
    resources 74; strategy factors **62**
Research Institute for Economy; Trade and
    Industry (RIETI) 55
Research proposition 56, 78, 161
Research Question 116
research questions 9, 31, 94, 140
resources access 163
restructuring 2
Rights to Compensation 77–92
risk averse 14
risk taking 14, 15, 25
risk taking orientation 20
Ritsumeikan University 166, 167
Rohm 163, 170
role of SMEs and Start-ups 69–73, 72
rule-regulated network 129

St Louis life science cluster 48
salarymen venture Capitalists 147
sales growth *38*
Samsung 103, 105
Samsung Electronics 10, 14
Samsung of South Korea 100
Schumpeterian types of companies 162
science and technology 2
Science and Technology Basic Plan 141
science and technology evolution 3
scientific infrastructure 2
Scientific seeds *34, 35*
seed capital 172
Sega 121, 126, 128; programme for start-ups 126
Sega (Farmicom Mega Drive) 122
Sega Saturn 122
SEMATECH semiconductor industrial project 159
semiconductor companies 111
semiconductor industries 14, 26, 179
Semiconductor Industry 93–114
Semiconductor Industry Research Institute Japan (SIRIJ) 108
Semiconductor Leading Edge Technologies (SELETE) 109
semiconductor manufacture 95
Semiconductor Technology Academic Research Center (STARC) 109
semiconductors 110
seniority value 20
service industries 162
Sharp 164
Shibuya; Tokyo 121
Shikoku 143
Shimosaka, S., President of Japan Patent Attorneys Association (JPAA) 83
Shitara; Judge 80
significance of NTBFs and UICs 55
Silicon Valley 121, 148
Silicon Valley model 122, 133, 158, 172, 177, 179
Silicon Valley-type companies 163
Singapore 26
Size of Firm 59–61
skill development 126
Slovak Republic 41
Small Business Administration (SBA) 38
Small Business Innovation Research (SBIRs) 42, 44
Small Business Tech Transfer Grants (STTRs) 42, 44
small and medium-sized enterprises (SMEs) 5, 6, 46, 56, 58, 60, 63, 65, 66, 69–73, 74, 78, 88, 118, 122, 158, 162, 165, 170
socio-economic factors shifting 96

sociopolitical cultures 48
Softbank 162
SOGO Clinical Pharmacology Co. Ltd. 170
SOIKEN Medical Science Study Laboratory 170
Sony 104, 105, 121, 122, 128, 163, 180
Sony Computer Entertainment (SCE) 122
South Korea 176
South Korea innovation 4
South Korean companies 158
specific industries specialization 13
sponsoring outsourcing competences 127
Square Enix 121
stagnation option 3
Stanford University 6, 32, 37, 43
start-ups 118
Status of the Patent Attorney (Benrishi) 82
stem cell research 48
stock exchanges 171
Storz, C. 6, 115–38, 162
Stowers Institute in Kansas City; Missouri 48
Stowers, J. 48
Stowers, V. 48
strategic alliances 26
strategic behaviour 13
Strategic Council on Intellectual Property 82, 83
strategic role of talent 178
Substantive Patent Law Treaty (SPLT) 87
Sun Microsystems 163
Super Famicon 122
Super Silicon Crystal Institute (SSi) 109
Supreme Court Olympus v. Tanaka 81
system-in-a-package (SiP) 103
system-on-the-chip (SoC) 103

Taiwan 26, 176
Taiwanese semiconductor companies 103
talent 177
Tama 171
Tamai, K. Professor 80
Taplin, R. 5, 77–92, 169
techno-governance concept 95
Techno-Governance structures 95–9, 96, 97, 100
Techno-governance structures and contingencies **98**
technological catch-up 22
technological collaboration 15
technological dynamics 97
technological innovation 95
technology: acquisition 15, **16**; alliances 16; diffusion 48; frontier projects 64; imported 11; and innovation 3; outsourcing 24; sources 22; sourcing 15; supporting organizations (TSOs) 94; transfers 165

Technology Licensing Organization (TLO) 4,
31, 34, 61, 78, 120, 160, 165; Law 34;
Promotion Law 55
Telecommunication Research Institute (TRI)
102
temporary hires (*arubaito*) 126
tertiary education system transformation 159
Texas and A&M University 79
Texas Instruments (TI) 105
timing of commercialisation **60**
TLO Laws 78
Tohoku University 32
Tokai University 167
Tokyo 143, 158, 164, 166
Tokyo District Court 90
Tokyo High Court 80, 90
Tokyo metropolitan 149
Tokyo University 6, 32, 37, 80, 160, 166
Toronto University 32
Toshib-Matsushita Display Technology 108
Toshiba 82, 104, 105
Toshiba-Fujitsu-Sony 108
total quality management 78
Toyota 14
Toyota Production System (TPS) 1, 176
Toyoto 129
trademark examination 85
Trademark Law Treaty (TLT) 88
trends in venture capital investment *43*
triple helix cooperation 177
Tsukagoshi, M. 6, 139–56
Tsukuba 1
Tsutaya 162

Unauthorized Copies 87
uncertainty 125
Uniqlo 162
United States of America (USA) 4, 30–53, 160,
163, 169, 176; business angels 145, 149;
Food and Drug Administration (FDA) 45;
Patent and Trademark Office 11
United States Patent and Trademark Office
(USPTO) 88
United States-Japanese semiconductor disputes
102
university: changing roles 2, 164; privatization
150; spin-offs (USOs); university
performanced 166; start-ups 31, 37;
system 165; system reform 61; technologies
170

University of California (UCLA) 32
university industry collaboration *67*; activities
55, 59–61; obstacles **61**
University Industry Collaborations
Activities/Research Productivity 65–8
university reform 45
university spin-offs (USOs) 6, 139–56, 152;
emerging role 159, 164; emerging tys 172;
needs and challenges 150; newly established
142; ownership and support 167; problems
154; role 157–74; seed/early-stage 151;
support programmes 153
University of Tokyo 143
university-based researchers 160
university-sponsored incubators 38
user-driven innovation 131

venture businesses 118
Venture Capital Association (NVCA) 40
Venture Capital (VC) 39–42, 172; activities 79;
funding 157; investment stages 40; investors
141; lack 120; markets 169; system 31
Venture Capitalists (VCs) 145; hands-on 145;
investment style 147
Venture Enterprise Center (VEC) 146
Venture incubation centres 153
Very Large Scale Integration (VLSI):
Cooperative 108; function-specific 100

Waseda 167
Waseda University 143, 166, 171
western patent law 77
Wii of Nintendo 122
Williamson, O. 96, 97
wireless Internet 129
Wisconsin Alumni Research Foundation
(WARF) 34
Withdrawal of Macro- and Micro-level
Strategies 102
World Intellectual Property Organization
(WIPO) 87

X-Box by Microsoft 122

Yawata, K. Chairman of IAI Japan 153
Yokoga Electric 103
Yonezawa, S. 81
Young, T. 79

zero defect 1